# So You Want to Be a Chef?

## Your Guide to Culinary Careers

## Second Edition

....................

## Lisa M. Brefere

CEC, AAC

## Karen Eich Drummond

Ed.D, RD, FADA, LDN, FMP

## Brad Barnes

CMC, CCA, AAC

....................

**WILEY**

**John Wiley & Sons, Inc.**

For general information on our other products and services, or technical support, please contact our Customer Care Department within the United States at 800-762-2974, outside the United States at 317-572-3993 or fax 317-572-4002.

Wiley also publishes its books in a variety of electronic formats. Some content that appears in print may not be available in electronic books.

For more information about Wiley products, visit our Web site at http://www.wiley.com.

**Library of Congress Cataloging-in-Publication Data:**

Brefere, Lisa M.
    So you want to be a chef? : your guide to culinary careers / Lisa M. Brefere, Karen Eich Drummond, Brad Barnes.—2nd ed.
        p.   cm.
    Includes index.
    ISBN 978-0-470-08856-2 (paper : alk. paper)
        1. Cooks—Vocational guidance.   2. Food service—Vocational guidance.   3. Cookery—Vocational guidance.   I. Drummond, Karen Eich.   II. Barnes, Brad.   III. Title.
    TX649.AIB748 2008
    641.5'023—dc22

                                                                            2007050141

Printed in the United States of America

10 9 8 7 6 5 4 3 2

I NEVER PROPERLY THANKED THE PEOPLE IN MY LIFE who helped me nurture my abilities, fuel my passions, and evaluate my flaws. I am grateful to have the opportunity to dedicate this book to those who have been part of my culinary journey. The impact of these people on my career can never be properly measured or reciprocated.

I dedicate this book, which celebrates the enthusiasm and energy of our profession, to:

**Ed Esposito,** for giving a young woman with short checked pants and red socks an opportunity to step confidently into a man's kitchen with respect and equality.

**Rene Mettler,** for recognizing my natural talent, teaching me the art of dedication, and meticulously refining my skills.

**Fritz Sonnenschmidt,** for mentoring my path, cultivating my vision, and always finding the time to talk.

**Brad Barnes,** my partner, my friend, for keeping the bar high, never compromising on quality, and for giving me the desire to strive for individual perfection.

**To my family:  My beloved mother,** for teaching me to balance my life, to persist in reaching for my dreams, and to remember the words "never give up." **My dad,** for my work ethic, my inventive mind, and my obsession for perfection. **My sister Leslie,** who fostered my childhood, teaching me love for family traditions, guiding my many choices, supporting my decisions, and giving countless hours of counseling. **My brother, Edward,** for urging me to stay focused, develop leadership qualities, and follow my dreams.

**My sister Lauren,** for cherishing her baby sister with the love and care of a mother and coaching me with a good-natured heart and a nonjudging character, giving me the aptitude to excel in my career. M**y children, Joey Jr., Julia, Johnny, and Jeremy,** for keeping me young at heart, whole in spirit, and openminded in thinking. And to **Joe, my husband,** with thanks for the decades of support, the years of encouragement, the months of advice, and the lifetime of friendship. Your level, calming nature has taught me patience and respect and showed me opportunities I never would have found on my own. You showed me how to take time to pick the tomatoes.

My inspiration for this book is a direct result of the wonderful people in my life. They have given me the strength and perseverance to encourage others following a similar career path. Thank you all for allowing me to give back.

LISA BREFERE, CEC, AAC

**THIS BOOK IS DEDICATED TO OUR INDUSTRY,** to the camaraderie of chefs and cooks. It is dedicated to the love for food we all share and our fascination with the effect heat has on ingredients. This book was written because our industry, although sincere and rewarding, can be an undescribed, directionless, intimidating maze, and over the years I have seen too many young cooks with fear and curiosity in their hearts about the uncertainty of just where this career will lead them. It is my greatest hope that this book will enlighten its audience, remove the mystery, and help all interested find their way through the fantastic journey of love, craft, art, and science we all take for granted as we enjoy each plate of food put in front of us.

Most of all, this book is dedicated to the brigade of incredibly dedicated people who have helped me see my way in my chosen career, the men and women I recognize as my mentors. I hope I have taken their teachings and formed them into a tool that will help many more culinarians than I could ever touch otherwise.

BRAD BARNES, CMC, CCA, AAC

# Contents

## Introduction ix

## PART 1  FEEDING FRENZY ....................................... 1

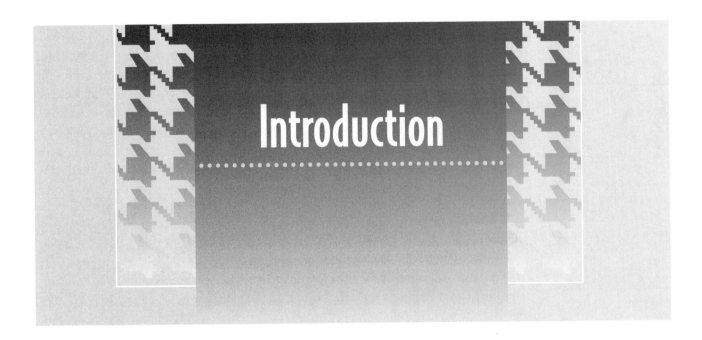

# Introduction

## Welcome

by Lisa Brefere, CEC, AAC

### "SO WHY DID YOU WANT TO BECOME A CHEF?"

After 27 years as a professional Culinarian and Chef, this is still the question I am most frequently asked. When I, a woman, picked this career 28 years ago, family, friends, and relatives dubbed me certifiably crazy. "Why would you want to be a Chef?" they asked. "To flip hamburgers the rest of your life?" Truthfully, no one thing lured me into this profession. I graduated from a well-known culinary school and come from an Italian-American background; thus, my career, and its inspiration, has centered on wholesome foods, natural flavors, and the freshest possible ingredients.

Early memories of my childhood revolve around large family meals, with wonderful smells of traditional dishes permeating the house. At least monthly we would get together with family and friends at a huge buffet table overflowing with delicacies like Aunt Columbia's crab corn bread, Mrs. Armando's orange butter cake, and Aunt Cindy's veal cutlet with olives and fresh garden tomatoes. I vividly remember coming home from nursery school and finding a big paper bag of something moving in the refrigerator. When I pushed the bag, something snapped back at me. That was my first experience with lobsters, the king of the sea, as my mother called them, when she taught me at the tender age of five the art of cleaning and cooking this treasure. It wasn't until some years later that I realized my early dining experiences were not the norm for the 1960s, when TV dinners, frozen vegetables, and convenience foods were becoming popular.

In 1975, I was introduced to the retail aspect of the culinary business at an obscure little store in Hayward, California. It was a drive-in dairy concept where customers drove into the middle of the store, where all the products were visible. The customers called out orders as I ran around collecting their groceries from a large selection of local dairy products, canned goods, homemade tortillas, bakery items, and fresh seasonal produce. I was 15 and thought the whole thing was fascinating. At the end of the day, the owner gave me expiring dairy, wilted produce, and day-old bread to take home, where I lived with my big sister. These I cooked, along with a limited amount of secondary cuts of meat and poultry. Although this was the most primitive phase of my career, it was the most innocent and adventurous. I didn't have a reputation to uphold; I just did whatever I wanted, and everyone in the house was thrilled to be catered to, as long as they didn't have to help or clean. I was so taken by the experience of cooking that I never minded the work, which was always followed by praise.

My next experience was at an old-fashioned coffee shop, gourmet restaurant, and airplane catering company located at the Westchester County Airport. This is where I was introduced to my first real kitchen arrangement, complete with a hot/cold station, expediter, meal period rushes, beautifully displayed foods catered to corporate jets, and the adrenaline buzz of a busy environment. There was a magic about the kitchen, a sense of passion, commitment, love, and enthusiasm. You are there as a Chef by choice, pursuing a profession that grows from within, an occupation that reflects history, tradition, honor, and pride. Either you love it or you hate it. The heat, smoke, work, tireless hours just go with the territory. It's magnetic, exciting, ever-changing, and rewarding. You get up every day and love what you know, what you do, and what you get paid for.

## DO YOU THINK YOU ARE BORN TO BE A CHEF?

Perhaps you are, perhaps not, but what I do know is you are born with work ethics, pride, dedication, and perseverance. You were born to lead, born to teach, born to mentor, and, most important, born to create. I didn't know what I wanted either. I loved the scene of the kitchen, the fast pace, the routine that called for immediate attention, instant results, split-second decisions, long-term goals, with many short-term conclusions. I knew early on I needed to work with my hands, but at the same time I needed to use my intelligence, to be a motivator. I was convinced from the beginning of my education that my hands would work in tandem with my mind, and cooking would be the driving force of both.

## SO YOU WANT TO BECOME A CHEF?

You don't become a Chef; you strive, you excel, you achieve, you commit, and you train. Your heart throbs, expediting a busy night, watching your creations sail out the door with your signature attached. I can't remember the most dramatic and wonderful experience in my career that validated my decision to become a chef. There have been so many over the years, and so many more are to come. I can select examples to reflect on that are not only highlights but events I am extremely proud of: my first job at Greenwich Country Club with Chef Edmund Esposito, who was the only chef in my area that would hire a "girl," allowing and supporting my initiative to become the first ACF Women apprentice; when as a young aspiring culinarian, at my first culinary competition, being critiqued by the great Herman Rush, when he gave me his dignified nod of approval; when Rene Mettler invited me to carve ice in Japan on the American Ice carving team; when Fritz Sonnenschmidt welcomed me to the 1988 Big Apple Chapter Olympic Team; and seven years ago, when my longtime colleague and friend, Brad Barnes, CMC, and I teamed up to open our own culinary concepts and consulting company, B&B Solutions.

There have been honors and accomplishments that both reinforced my choice and reenergized my desire to be a chef.

The motivation for *So You Want to Be a Chef? Second Edition* is to share the journey of a profession and the wisdom, knowledge, and guidance of today's industry leaders. As a new culinarian, it is essential to map a career path, to understand your many choices, so you avoid taking unnecessary turns that hinder your success. For this second edition, we decided to divide the original book into two versions. *So You Want to Be a Chef? Second Edition* identifies the many varied choices of occupational paths for a Chef, with realistic vivid descriptions of what a day in the life would encompass. We have included interviews from industry leaders who have become successful in some of these choices. There are quotes from recent graduates from a variety of culinary programs that succinctly capture their advice to current culinary students. The companion book, *So You Are a Chef,* focuses on your career planning, resume writing, and building a portfolio toward the end of, or after the completing of, your formal education and vocational start.

Take a moment to read the fine print instead of jumping into a decision that can cost you time and valuable experiences in the end. We present you with a lifetime of knowledge and expertise to help you advance into a successful career with many necessary tools. You have chosen a wonderfully rewarding field that can take you to places you never imagined possible. Now, keep these fundamentals in your pocket for the passage that lies ahead: Listen. Remember. Write Down. Ask Questions. Practice. Learn the Basics. Develop a Style. Stay Humble. Keep Listening. Continue Your Education. And most important: Give Back to Others.

In Food we Trust!

## Welcome
### by Brad Barnes, CMC, CCA, AAC

It really all started with a love for food. At a very young age, I remember being amazed that the things grownups called fruits could be so incredibly tasty and even grow on trees. The things we could eat were infinite and all had a life of their own; nothing tasted or felt quite like anything else. One of my earliest gourmet meal experiences was at the Officer's Club in Fort Benning, Georgia, at the age of five, when I dined on escargot and frogs' legs. It was of course not until later that I realized, in my grandmother's kitchen, that there was a real plan, a special way to do things, so that food might taste so good. I was puzzled and intrigued by the way hot oil made meat or potatoes turn brown, and that brown by itself tasted different from the inside of the food. I wondered why my grandmother put milk on top of the fridge with old biscuits and corn bread. She told me so the milk would curdle and the bread would get hard. But then she mixed it all up in a couple of days with sugar and eggs and baked it into an indescribably delicious product she called bread pudding. It all seemed quite magical.

I don't think I ever really accepted that it was easy to make food taste good; it seemed to be so cared for by anyone cooking. Cooking or "fixing" a meal appeared to be a process and always done in very specific ways, for each item to be prepared. Of course I now know that the methodology behind this application of heat to edible goods in order to render them palatable is far more complicated than my childhood mind could have ever guessed. But I felt it was all in the way it was done, and not much else could affect the outcome. This notion was reinforced by watching Julia Child and the Galloping Gourmet along with the Three Stooges or Bugs Bunny, all of which had a profound influence on my being as a chef and person. By my early teens I was preparing foods myself and trying to refine the methods I used to control the product's behavior. I read different sources on the same subject, drew my conclusions, and made my own adjustments to the printed cooking methods.

When I needed to get a job, I discovered I could actually be paid to cook! So I got a job at the age of 15 in a high-volume Italian-style restaurant. I was in charge of the salad bar for a bit, and then I moved through the kitchen at each station, loving every step. When I went off to college, my need for money increased, and so did my interest in cooking. I secured a job as a pantry cook and again moved through the stations. At some point (this is not a career path recommendation), I decided that cooking was much better than college, and I quit school. Inside of a year, I had become a reasonably integral part of a restaurant team and was asked to help open a new project in the suburbs. I started as a supervisor and within six months the Chef and Sous had bolted and I was made Chef. Talk about a rude awakening! I toiled as never before to succeed at this task, and stayed for five years.

At this point I attained my culinary education, developed goals, and began to move forward with my career. I became a competitor and began to garner medals at competitions, with my sights on the Culinary Olympics in Germany. I went, won perfect gold, and continued working toward my goal of becoming a Certified Master Chef, which I attained in 1996. Another eight years have passed, along with three more Olympics, and I sit writing to you about the fantastic trip I have experienced, all in the name of food.

I must convey to you that *So You Want to Be a Chef? Second Edition* was primarily written to outline the many possibilities in our industry and to be an aid in planning your career. The motivation to create a book with this type of message is to help the young aspiring culinarians of today avoid making career mistakes that can take years to mend. The goal is to mentor your success, so why guess at the different choices when we can assist in achieving this mission? To have the ability to give back after many years of experience is one of the most powerful gifts one can accomplish. You should realize, however, that no matter what, to be really good, there is a secret requirement, a special skill you will need, an ability held dear by those who possess it: You must be a good cook.

I do not want to try to quantify why this business held my heart so strongly, but I can tell you that few things in this life have grasped my soul as profoundly as the knowledge of how to make people (and myself) feel good through the art of cooking. Of course, the science involved is intriguing, and the art of the process and all its infinite parts form a never-ending challenge, but I still think that the people we affect are the true reward of mastering this craft. When you can gather a group of friends to enjoy your work or serve someone confined to a hospital bed whose only pleasure is your meals, then you know what the real satisfaction of this business is. To demonstrate to a class of young cooks the way egg whites become the body of a perfectly risen soufflé, or how the roux changes to a nutty brown before it's used to build a silky espagnole sauce, will help you realize how precious our craft is. As you watch the bride and groom sit in their special places enjoying your heartfelt labor and realize you are a big part of so many special times in so many people's lives, then you will know what you have done with your life. And as you sit back and remember the fragrance of your grandmother's home just before dinner, when all you did was play and eat, or the smell of break-fast when you rolled out of bed, it will truly be undeniable that to be a good cook you must be one part artist, one part scientist, and one part magician.

# The Culinary Professional

People who work as Chefs and in other positions in the culinary field are considered culinary professionals. A professional is someone who works in a specific type of occupation referred to as a profession. Fields such as teaching and nursing are also referred to as professions. A profession has the following characteristics.

- It requires mastery of a specialized body of knowledge and an extensive period of training.
- It provides a service to society.
- It maintains high standards of conduct and competency for its members.
- It offers its members opportunities to become certified or licensed and to be active in professional organizations.
- It asks members to be committed to lifelong learning or continued study.

As you can see in Figure I-1, the culinary profession meets these requirements by being service-oriented and having a body of knowledge, code of conduct (Table I-1, from the American Culinary Federation), certification programs, and lifelong learning.

**Figure I-1**
The Culinary Profession

Adopted by the American Culinary Federation, Inc., at its national convention in Chicago, August 1957. Copyright © 2004 American Culinary Federation. All rights reserved. Reprinted with permission.

The American Culinary Federation (ACF) is the largest and most prestigious organization dedicated to professional chefs in the United States today. Their mission is as follows: "It is our goal to make a positive difference for culinarians through education, apprenticeship, and certification, while creating a fraternal bond of respect and integrity among culinarians everywhere."

Getting certified in the culinary field is a tremendous asset and can help you advance your career. ACF offers the only comprehensive certification program for Chefs in the United States. ACF certification is a valuable credential awarded to Cooks, Chefs, Pastry Cooks, and Pastry Chefs after a rigorous evaluation of industrial experience, professional education, and detailed testing.

ACF also offers its members many opportunities to keep up-to-date with the latest in knowledge and skills through its monthly publications (*The National Culinary Review* and *Center of the Plate*), seminars, workshops, national conventions, and regional conferences. ACF sanctions U.S. culinary competitions and oversees international competitions that take place in the United States. ACF also accredits culinary programs at the secondary and postsecondary levels. Local chapters of ACF offer members opportunities to network with nearby culinary professionals. The ACF chef apprenticeship program is described in the next section.

Culinary professionals work in the foodservice industry, which is a part of the hospitality industry. Hospitality can be defined as giving guests a generous and cordial reception, and then offering them a pleasant and sustaining environment. The hospitality industry includes hotels, restaurants, and any other kind of institution that offers shelter, food, and often additional services to people away from their homes. The hospitality professions are among the oldest in the world, for cooks and innkeepers alike have a long history of making guests feel at home. Inns and taverns have played an important role in history from the ancient Greek taverns to the European inns during the Crusades, the old English inns, and eventually the early American taverns, inns, and hotels.

# Culinary Training Programs

Formal training in the culinary field is imperative for anyone who wants to work as a chef. In the past, some chefs learned their trade through long-term apprenticeships. However, today's professional culinary programs teach a lot more than just cooking and baking. Most culinary programs now offer more business courses (financial management, marketing, and so on), as well as computer training, to better prepare chefs to assume leadership and managerial roles in the industry and to manage large, complex foodservice operations. Culinary training also has adapted to reflect changing food trends and eating habits. For example, chefs and cooks must know a wide variety of food preparation techniques and cooking styles. They also must know how to prepare foods to accommodate various dietary restrictions, to satisfy health-conscious eating styles and to meet the needs of an increasingly international clientele. Chefs and cooks also must be creative and know how to inspire other kitchen staff to develop new dishes and inventive recipes.

Professional culinary courses are available at high schools, culinary schools, two-year colleges, four-year colleges and universities, hotel and restaurant associations, and trade unions. The Armed Forces are another source of training and experience in foodservice work. Satisfactory completion of coursework can result in the award of a certificate or diploma, depending on the program.

Certificate programs usually last from several weeks to as long as 12 or 18 months. These programs offer certificates (not degrees) upon completion. Many types of certificate programs are available for individuals looking to start a career in the culinary field, and also for experienced culinary professionals looking to expand and update their skills. You can find certificate programs at many schools, colleges, and universities.

Culinary degree programs at culinary schools, colleges, and universities usually offer associate degrees, but more are also offering bachelor's degrees. An associate degree requires two academic years of study. The most common associate degree for the culinary field is the associate of occupational science (AOS) degree. Other associate degrees include the associate of arts (AA), associate of science (AS), and associate of applied science (AAS).

Obtaining a bachelor's degree requires four years of study. Depending on the program, culinary students pursuing a bachelor's degree take courses in management, marketing, finance, communications, purchasing, and cultural cooking as well as cooking and baking. If you plan to complete an associate degree first, and then pursue a bachelor's degree at another college, chances are that some, but not all, of your credits for your associate degree will transfer to your bachelor's degree. Try to find out early which credits are transferable.

Many culinary schools and colleges operate year-round, admitting students every few weeks. To apply, you must have a high-school diploma and, sometimes, standardized test scores (SAT or ACT) and work experience (though not necessarily in cooking). Many schools have three culinary programs: culinary skills, baking and pastry, and management. Students often combine programs.

To find culinary schools in your area, you can use the Internet and the Yellow Pages. A number of online sites list culinary schools, but the only site that lists all possible culinary schools is a federal website, http://nces.ed.gov/ipeds/cool/. The Integrated Postsecondary Education Data System (IPEDS) College Opportunities On-Line is your direct link to nearly 7,000 schools, colleges, and universities in the United States. This website is sponsored by the National Center for Education Statistics in the U.S. Department of Education, and it is designed to help college students and future students understand the differences between schools and colleges, and how much it costs to attend them. You can find out about a specific school, if you have one in mind, or you can search for a school based on location, program, or degree offerings. On this site, you can search specifically for Personal and Culinary Services programs.

Another way to become trained as a Chef is through an apprenticeship program. The ACF sponsors more than 80 culinary apprenticeship programs in the United States. ACF apprentices are required to work for three years under the supervision of a Chef and to attend school part-time throughout the apprenticeship. Apprentices gain work experience in all the kitchen stations and learn about recipe and menu development, purchasing, financial management, sanitation, and safety. Information about ACF apprenticeships is on their website (www.acfchefs.org).

Table I-2 lists questions to consider when choosing a culinary program. One of the questions to ask is whether or not the program is accredited. Accreditation is a status granted to an educational institution or a program found to meet or exceed written standards of educational quality. In the United States, accreditation is not a requirement, but most schools (beyond high school) are accredited or in the process of becoming accredited. Accreditation ensures that the school or program offers a high-quality education to its students.

The Council for Higher Education Accreditation (CHEA) recognizes the following regional accrediting agencies for colleges and universities in the United States.

◆ Middle States Association of Colleges and Schools
◆ New England Association of Schools and Colleges
◆ North Central Association of Colleges and Schools
◆ Northwest Association of Schools and Colleges
◆ Southern Association of Colleges and Schools
◆ Western Association of Schools and Colleges

In addition to these regional accrediting agencies, several national agencies are active, including the American Culinary Federation Educational Institute Accrediting Commission. Accreditation by ACF requires that curriculum, faculty, facilities, resources, support staff, and organizational structure all substantially meet the standards set by the Accrediting Commission. These standards were set by industry leaders in both culinary arts and culinary arts education. They are monitored

## Table I-2 Questions to Consider When Choosing a Culinary School

**Is the School in a Convenient Location?**
Do you like the campus?
When are classes scheduled? Does the program offer a flexible schedule: day, evening, and on-line options?
What are the admission requirements?
What courses are required to complete the program?
How long will the program take to complete?
Do the students like the program?
Are the kitchens and other facilities sufficient and up-to-date?
Does the school run a restaurant, bakery, or similar operation?
Does the school offer internships or externships where you work in a kitchen for credit?
Will you get a certificate or a degree? If you get a degree, what type of degree?
Is the school accredited?
Do the instructors seem knowledgeable and approachable?

What are the qualifications of the faculty?
What is the instructor-student ratio?
How many contact hours do the students receive in the kitchen labs or other lab components of the program?
What is the job placement office like, and what are their job placement rates? What is the average starting salary for graduates?
What types of financial aid are available?
How does the school select financial aid recipients and determine your financial need?
How much debt may you incur to attend school, and how does it compare to what you expect to earn once you are working?
What percentage of students complete the school's programs? What percentage transfer out?
Who are the alumni, and what kind of positions do they hold in the industry?

constantly by the Accrediting Commission for accuracy in the culinary profession. The standards are the centerpiece of what constitutes a viable program.

Culinary or foodservice programs that are ACF-accredited have been judged against the published standards. They undergo a self-evaluation and report their findings to the Accrediting Commission. The commission then authorizes a fact-finding team to visit the school to verify compliance to standards. The self-evaluation, the report of the fact-finding team, and the program response are studied by the commission, which then grants accreditation to deserving programs. Check the ACF website (www.acfchefs.org) to see if the program you are interested in is accredited by ACF. Another national agency is the Accreditation Commission for Programs in Hospitality Administration (ACPHA). ACPHA is the accrediting arm of the Council on Hotel, Restaurant, and Institutional Education (CHRIE).

Once you have started in a culinary program, make the most of it by working hard. You can also start your career off right by doing any of the following while you are in school:

- Get involved as a student member of ACF.
- Join any other culinary student organization and become an active member.
- Enter an ACF-sponsored student culinary competition.
- Go to local trade shows and conferences. Students can usually attend for a greatly reduced fee.
- Get work experience in the field, either as part of an externship/internship or on your own.
- Start building a portfolio with photographs, certificates, class projects, recipes that you have tested, and so on. Document everything you do that could be useful when you are looking for a job or internship.

## Earnings and Outlook

In general, job openings for Chefs and Cooks are expected to be plentiful through 2014. Overall employment of chefs, cooks, and food preparation workers is expected to increase about as fast as the average for all occupations between now and 2014. Employment growth will be spurred by increases in population, household income, and leisure time that will allow people to dine out and take vacations more often. In addition, the large number of two-income households will lead more families to opt for the convenience of dining out.

Projected employment growth, however, varies by specialty. The number of higher-skilled chefs and cooks working in full-service restaurants—those that offer table service and more varied menus—is expected to increase about as fast as the average. Much of the increase in this segment, however, will come from job growth in more casual dining, rather than upscale full-service restaurants. Dining trends suggest increasing numbers of meals eaten away from home and growth in family dining restaurants, but greater limits on expense-account meals.

**Table I-3 2006 Mean (Average) Hourly and Annual Wages for Chefs and Head Cooks**

| | INDUSTRIES | |
| --- | --- | --- |
| | Mean (Average) Hourly Wage | Mean (Average) Annual Wage |
| All Industries | $18.21 | $37,880 |
| Full-Service Restaurants | $17.15 | $35,670 |
| Hotels and Lodging | $21.19 | $44,080 |
| Cruises | $18.96 | $39,430 |
| Elementary and Secondary Schools | $16.49 | $34,300 |
| Colleges, Universities, and Professional Schools | $20.19 | $42,000 |
| Grocery Stores | $18.40 | $38,270 |
| Hospitals | $20.23 | $42,090 |
| Nursing Care Facilities | $19.15 | $39,830 |

| Top-Paying States for Chefs and Head Cooks | Top-Paying Metropolitan Areas for Chefs and Head Cooks |
| --- | --- |
| District of Columbia | Nassau, Suffolk Counties, New York |
| New Jersey | Edison, New Jersey area |
| New York | Decatur, Illinois |
| Hawaii | Sebastian, Vero Beach, Florida |
| Nevada | Bridgeport, Stamford, Norwalk, Connecticut |

Source: U.S. Department of Labor, Bureau of Labor Statistics, 2006 National Occupational Employment and Wage Estimates.

Chef salaries vary greatly according to region of the country and type of food-service establishment. Salaries are usually highest in elegant restaurants and hotels, where many Executive Chefs are employed, and in major metropolitan areas. Salaries also rise as the sales volume of the operation increases. Table I-3 shows the average hourly and annual wage for Chefs and Head Cooks in various industries and locations as of 2006.

## What's New to This Edition

The biggest change to *So You Want to Be a Chef? Second Edition* is that the section on managing your career has been removed and made into a new companion book,

*So You Are a Chef.* As a result, *So You Want to Be a Chef? Second Edition* now enables the culinary student or someone interested in pursuing a culinary career to concentrate on the varied choices available in the industry. *So You Are a Chef* is the next step in managing your career, whether you are soon to graduate from a culinary program, have recently graduated, or have been working in the industry and are looking to make a change. Other changes to this edition include:

♦ A new chapter on Research and Development Chefs, an exciting emerging field in the culinary world.

♦ Advice from recent culinary school graduates to current culinary school students on the advantages of their education.

♦ Chapter 9 has been broadened to cover careers in on-site foodservice, now highlighting careers in health care, continuing care communities, the armed forces, and more.

♦ The integration of career path advice throughout the book, rather than only at the beginning of the book.

♦ In order to streamline each chapter, professional organizations are identified in each chapter, but the details of the organizations can now be found in Appendix A.

## How to Use *So You Want To Be a Chef? Second Edition*

*So You Want to Be a Chef? Second Edition,* is organized in three parts, as shown here. The book discusses culinary careers in over 20 industry segments.

### PART 1. FEEDING FRENZY (CHAPTERS 1–6)

These chapters discuss culinary careers in mostly traditional and always busy settings, including hotels, restaurants, cruise lines, clubs, catered events, and supermarkets. At the beginning of Part 1 is a discussion of possible career paths and education path advice.

### PART 2. FEEDING THE MASSES (CHAPTERS 7–9)

These chapters tell you about careers in on-site foodservices such as business and industry, schools and universities, hospitals, retirement communities, and the military. In on-site foodservices, the foodservice is offered to employees, students, patients, residents, and the military as a needed service. At the beginning of Part 2 is a discussion of managed service companies, career paths, and education path advice.

### PART 3. HOME ON AND OFF THE RANGE (CHAPTERS 10–12)

The three chapters in Part 3 discuss culinary careers that may or may not involve as much cooking as found in the jobs discussed in Parts 1 and 2. New to this section is a dedicated chapter on the career path for a Research and Development Chef. Learn about working as a Private or Personal Chef, Culinary Educator, Purchasing Director, Food Stylist or Food Photographer, Food Editor, Food Writer and Author, Public Relations Manager, and even a Celebrity Chef.

To help you get the most out of Chapters 1 to 12, which discuss culinary careers in many settings (also referred to as segments), most chapters are organized using seven headings: The Feel, A Day in the Life, The Reality, Professional Organizations, Interview, and Organization.

## THE FEEL

This section is designed to give you the big picture: who your customers are, how you serve them, whom you work with. This section is written, like everything else in *So You Want to Be a Chef? Second Edition,* by people who have firsthand experience.

## A DAY IN THE LIFE

Ever wonder what a typical day is like for a Chef? Well, first of all, "typical" days are rare! Every day brings different customers, new situations, unique challenges. Even so, you will read firsthand accounts of Chefs describing how they interact with others, do their work, handle problems, enjoy the camaraderie of their coworkers, and more.

## THE REALITY

Being a Chef has its challenges, and this is where we talk about them in a no-nonsense way. You know the kitchen can often be hot and steamy, the work environment like a pressure cooker, and the workday many hours long. But we also talk about the unique realities of each segment: dealing with the executive committee in a hotel, working with hospital patients, and feeding troops in a war zone.

## PROFESSIONAL ORGANIZATIONS

In addition to the ACF, there are many other culinary professional associations. Table I-4 lists over 35 more associations with their website addresses. For more specific information on each association, see Appendix A.

## INTERVIEW

For each segment of the foodservice industry, you can read an interview with a seasoned professional who describes his or her job, work experience, and career path. They give you advice on the many ways you can achieve the goals of being a successful businessperson in the foodservice industry today. These interviews will give you insight into their real-life experiences—their many obstacles and

## Table I-4 Professional Associations

**American Culinary Federation (ACF), www.acfchefs.org**
American Dietetic Association (ADA), www.eatright.org
American Hotel & Lodging Association (AH&LA), www.ahla.com
American Institute of Baking (AIB), www.aibonline.org
American Institute of Wine and Food (AIWF), www.aiwf.org
American Personal and Private Chef Association (APPCA), www.personalchef.com
American Society for Healthcare Food Service Administrators (ASHFSA), www.ashfsa.org
Association for Career and Technical Education (ACTE), www.acteonline.org
Black Culinarian Alliance (BCA), www.blackculinarians.com
Bread Baker's Guild of America, www.bbga.org
Club Managers Association of America (CMAA), www.cmaa.org
Confrérie de la Chaîne des Rôtisseurs, www.chaineus.org
Cruise Lines International Association (CLIA), www.cruising.org
Dietary Managers Association (DMA), www.dmaonline.org
Foodservice Consultants Society International (FCSI), www.fcsi.org
Foodservice Educators Network International (FENI), www.feni.org
Grocery Manufacturers Association/Food Products Association (GMA/FPA), www.gmaonline.org/
Institute of Food Technologists (IFT), www.ift.org
International Association of Culinary Professionals (IACP), www.iacp.com
International Caterers Association (ICA), www.icacater.org
International Council on Hotel, Restaurant and Institutional Education (I-CHRIE), www.chrie.org

International Flight Services Association (IFSA), www.ifsanet.com
International Food Service Executives Association (IFSEA), www.ifsea.com
International Foodservice Manufacturers Association (IFMA), www.ifmaworld.com
Les Dames d'Escoffier International (LDEI), www.ldei.org
Military Hospitality Alliance (MHA), www.ifsea.com/military_inside.cfm?catid=4357
National Association of College & University Food Services (NACUFS), www.nacufs.org
National Association for the Specialty Food Trade (NASFT), www.fancyfoodshows.com
National Ice Carving Association (NICA), www.nica.org
National Restaurant Association, www.restaurant.org
National Society for Healthcare Foodservice Management (HFM), www.hfm.org
North American Association of Food Equipment Manufacturers (NAFEM), www.nafem.org
Research Chefs Association (RCA), www.culinology.com
Retail Bakers of America (RBA), www.rbanet.com
School Nutrition Association (SNA), www.schoolnutrition.org
Société Culinaire Philanthropique, www.societeculinaire.com
Society for Foodservice Management (SFM), www.sfm-online.org
United States Personal Chef Association (USPCA), www.uspca.com
Women Chefs and Restaurateurs (WCR), www.womenchefs.org
Women's Foodservice Forum (WFF), www.womensfoodserviceforum.com

successes, and the direction they used to achieve their career goals. These interviews should be used as a tool to let you identify your objectives for your education and employment opportunities. Use this knowledge as a guide to avoid unnecessary turns that will hinder your success and waste precious time in a career or education path.

## ORGANIZATION

Each chapter includes an example of an organizational chart. Actual job descriptions of various culinary positions are available in Appendix B. Real job descriptions can help you see what a chef does, as well as requirements for various culinary jobs.

## EDUCATION PATH ADVICE

Finally, most chapters contain the Education Path Advice guide. This guide displays the level of importance (through the height of the balls in the diagram) of formal education, knowledge, and competencies needed for positions in different settings. For example, in the chapter on hotels, the Education Path Advice will show you different levels of importance for classical cuisine, nutritional cooking, or a Masters Degree for various positions such as sous chef, restaurant chef, or executive chef. Education Path Advice is offered in Chapters 1–8.

## ADVICE FROM A RECENT GRADUATE

New to *So You Want to Be a Chef? Second Edition* is advice from recent culinary program graduates currently working in the profession. They provide their insight as to what they consider to be some of the most important information current culinary students can take from their learning environment.

# Resources for Instructors

To assist the instructor, an Instructor's Manual (978-0-470-25395-3) has been developed. It includes learning objectives, a chapter outline, class activities, and test questions for each chapter in *So You Want to Be a Chef? Second Edition.* In addition, PowerPoint® slides and student worksheets are available for downloading at www.wiley.com/college/brefere.

# Acknowledgments

We would like to thank the following people, who are not only consummate professionals, but close friends who have supported us in our journey in creating *So You Want to Be a Chef? Second Edition.*

**Roland Schaeffer, C.E.C, A.A.C., Retired Corporate Chef of Heinz Foods**
For his years of knowledge and his insights into the manufacturing, research, and development fields of the foodservice profession.

**Steve Goldstein, President, Food/Thinque**

For his marketing, sales, and food-marketing input in Part 3, "Home On and Off the Range." His contribution to the industry has been a driving force in the field of research and development.

**David St. John-Grubb, C.E.C., C.C.E., C.-I.E., A.A.C., Le Cordon Bleu, Miami**

For his charming comments on cruise line Chefs, which gave us the ability to express in writing the journey of a young apprentice aboard a grand vessel.

We are grateful to all of the interviewees for giving their time, expertise, and patience in presenting to you a taste of their career and education paths. Their commitment to our profession and their desire to give back what they have learned is an inspiration to all.

We also appreciate the feedback of the reviewers who read the initial proposal and subsequent drafts of the manuscript for both editions. They are:

John M. Courtney, Johnson County Community College

Emma Cutler-Grimes, New England Culinary Institute

Jamie M. Gizinski, Walnut Hill College

Sarah Gorham, The Art Institute of Atlanta

Joseph Renfroe, York Technical Institute

Odette Smith-Ransome, The Art Institute of Pittsburgh

David Weir, Orlando Culinary Academy/Le Cordon Bleu

# Part 1

# FEEDING FRENZY

Here's a glimpse of what to expect in Part 1, "Feeding Frenzy."

Just 20 minutes to showtime. You've been cooking and tasting all day in your restaurant. You scan a quick panorama of your kitchen: Production is cooled and put away; cook's mise en place filled, ready, and displaying their magnificent natural colors; the cooks are hanging, telling war stories of other busy nights; sauté pans are neatly stacked like a Doric column only an arm's length away from the stove; refrigerators stocked with seasoned proteins awaiting the first click of the POS; preparations on the cold station lined up perfectly like soldiers in battle formation waiting their first command. It's a feeling you must live, a feeling mixed with pride, obsession, perfection, compulsive actions that delight your guest with satisfaction. We strive to be the best we can, and we try damn hard.

Part 1 of this book discusses culinary careers and what working as a chef is like in:

1. Restaurants
2. Hotels
3. Clubs
4. Catered events
5. Supermarkets

For many chefs, these foodservices are the heart of the industry. They are busy, noisy, unpredictable, and full of activity. Being a chef in any of these venues is a passion, an art, an obsession that must be preserved and nurtured daily.

## FEEDING FRENZY CAREER PATH GUIDE

**Figure P1-1**
Feeding Frenzy
Career Path

Figure P1-1 is the Career Path Guide for culinary positions in restaurants, hotels, cruise lines, clubs, catering, and supermarkets. It is only a guide to give you an idea of what type of work experience you need, to move forward in your career. The Career Path Guide indicates the number of months in various positions that

### On the Way

Nonmanagement positions*

| Position | Executive Chef Hotel | Cutting-Edge Chef | Executive Chef Catering | Corporate Chef Multi-Unit Dining | Executive Chef Supermarket | Executive Chef Country Club | Executive Chef Cruise Ship |
|---|---|---|---|---|---|---|---|
| Prep cook | 8 | 12 | 12 | 8 | 8 | 8 | 8 |
| Pantry cook | 6 | 6 | 6 | 6 | 6 | 6 | 6 |
| Grill cook | 12 | 12 | 12 | 12 | 12 | 12 | 12 |
| Sauté cook | 12 | 12 | 12 | 12 | 12 | 12 | 12 |
| Expediter | 12 | 12 | 12 | 12 | 12 | 12 | 12 |
| Breakfast cook | 6 | 4 | 4 | | 8 | 8 | 6 |
| Banquet cook | 12 | 8 | 24 | 8 | 12 | 12 | 12 |
| Butcher | 4 | 12 | 6 | 6 | 6 | 6 | 6 |
| Roundsman | 8 | 8 | 8 | 12 | 8 | 8 | 8 |
| Retail counter cook | | | 8 | | 12 | | |
| Garde manger | 12 | 12 | 8 | 8 | 12 | 12 | 12 |
| Pastry cook | 8 | 12 | 8 | 8 | 12 | 8 | 12 |
| F. H. service staff | 8 | 8 | 8 | 8 | | 8 | 8 |

Management positions**

| Position | Executive Chef Hotel | Cutting-Edge Chef | Executive Chef Catering | Corporate Chef Multi-Unit Dining | Executive Chef Supermarket | Executive Chef Country Club | Executive Chef Cruise Ship |
|---|---|---|---|---|---|---|---|
| Off-premises chef | | | 12 | | | | |
| Sous chef | 12 | 24 | 12 | 12 | 12 | 12 | 6 |
| Banquet chef | 12 | | 12 | 8 | 12 | 12 | 12 |
| Restaurant chef | 12 | | | 24 | 12 | 12 | 12 |
| Executive sous | 24 | | 12 | | | 12 | 12 |

| Finish as | Executive Chef Hotel | Cutting-Edge Chef | Executive Chef Catering | Corporate Chef Multi-Unit Dining | Executive Chef Supermarket | Executive Chef Country Club | Executive Chef Cruise Ship |
|---|---|---|---|---|---|---|---|

*Number of months in the position at 40 hours a week.
**Number of months in the position at 55 hours a week.

ideally will prepare you for an Executive Chef position and others noted at the bottom of the chart.

The Career Path Guide will help you understand the time commitment needed to reach many of the top management positions in the foodservice industry. By no means do we claim that these paths are the only way, or that the time frames are precise. Some of you will advance much more quickly than we indicate, and some will need more time. The length of your workweek, your dedication, and your commitment are all factors in the time frame needed to obtain your final goal. Also, you may be able to hold two positions at once. For example, you could probably get experience as an expediter while working as a grill, sauté, or pantry cook. Similarly, you could be a roundsman, butcher, and *garde-manger* at the same time.

## FEEDING FRENZY EDUCATION PATH ADVICE

Within most chapters, you will find an Education Path Advice guide. The Education Path Advice guide displays the level of importance (through the height of the balls) of formal education, knowledge, and competencies needed for positions in different settings.

When you plan your educational path, the key is to remember that education is an investment in the future on which you will draw over your entire career. You may want to go to a two-year intensive cooking degree program, then work in the kitchen to perfect your skills, and return to school for another two years or so to get a bachelor's degree. Make a plan to maximize your time and your education.

The foodservice, hospitality, and culinary professions are first and foremost hands-on professions, but you should understand that they present minds-on challenges as well. As your hands carry out the precision tasks of common kitchen activities, your mind must retain, compute, and comprehend these important skills you are perfecting. As your mind absorbs the academic aspects of your profession—for example, financial reporting, computer techniques, menu costing, and menu development—the hands-on aspects make these skills come alive. The hands-on experience, which is the passionate and creative part of your profession, complements and motivates all facets of the work. Did you ever notice that when you are committed to a topic, you find that learning about it is easy? You're not studying, you're absorbing; you're not being tested, you're remembering.

Don't be intimidated by the Education Path Advice. Keep in mind that any academic program you need to complete will be filled with examples and situations related to the hospitality field that is your first love. Take one of the subjects in the education path that may sound threatening. For instance, food and beverage (F&B) financials may be scary because of the word financials, which makes it sound as if the class is all math problems. This is true—but the math problems all represent real-life situations you will need to know how to handle as your career unfolds. Financials and math problems are not as scary when you encounter them in the business you are committed to and passionate about.

Continue to advance your abilities with classes that will be an asset to your career. You could learn Spanish, which is spoken every day in many kitchens. Or you could take a public speaking course to help you do cooking demonstrations or other group presentations. These investments in your future will definitely pay off.

The educational tools we have mapped out are a matrix of the many directions available. We hope you use them to inspire, plan, and evaluate a successful journey toward a rewarding culinary career.

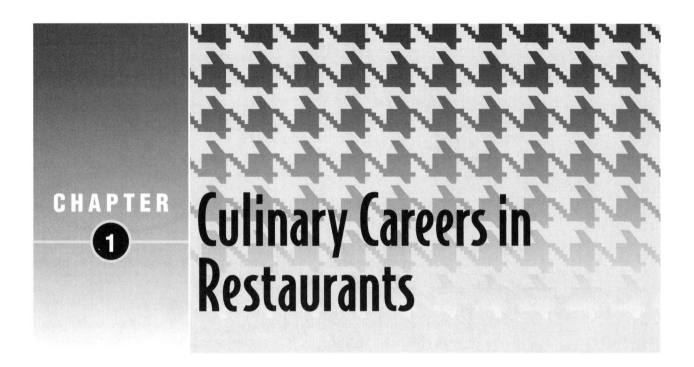

CHAPTER
1

# Culinary Careers in Restaurants

## Introduction

**NOT LONG AGO, EATING OUT WAS RESERVED** for special occasions and celebrations. Times have changed. With more dual-income families, fast and convenient meals are a must. Restaurants are now an essential part of the American lifestyle, with many Americans spending 47 cents out of every food dollar to dine out. With an annual economic impact of more than $1.3 trillion in 2007, the restaurant industry is huge, and it employs over 12 million people—more than any other private-sector industry.

The restaurant industry boasts over 900,000 restaurants in large cities, small towns, rural areas, and every place in between, and it presents consumers with more menu choices than ever before. About 45% of restaurants are limited-service eating places, such as fast-food restaurants and cafeterias, which primarily serve guests who select items and pay before eating. Full-service restaurants account for about 39%; these cater to patrons who order and are served while seated, then pay after eating.

National chains are a growing segment of full-service restaurants. These restaurants usually offer efficient table service, well-priced familiar menu items prepared by moderately skilled culinary employees, and a substantially nicer physical setting than limited-service establishments. By contrast, customers at upscale or cutting-edge dining places tend to seek a relaxed and elegant atmosphere in which to enjoy skillfully prepared food and leisurely but professional service. Cost-conscious and time-strapped guests increasingly eat at midscale or family restaurants rather than at elegant dining establishments.

You can categorize restaurants by level of service, such as limited service or full service. You can also use two culinary perspectives to roughly group restaurants. First, you can examine how much cooking is done from scratch and how much cooking utilizes convenience or premade products such as soup bases or frozen prepared foods. Second, you can look at who develops the menu and recipes. Are the menus and recipes mostly the product of corporate decisions, or is a chef solely responsible for menus and recipes?

Using the culinary perspective, most restaurants fall into one of these four categories:

1. **Cutting-edge restaurants:** Also called fine dining, these restaurants feature cooking from scratch almost exclusively, with the chef/owner developing the menus and recipes.

2. **Upscale casual restaurants:** These restaurants make most of their own menu items, and the chefs make most of their own menus and recipes.

3. **Value-driven, multi-unit restaurants:** The food in these restaurants is the result of corporate decision making and recipes, but it is still cooked mostly from scratch and of high quality. Examples include Houston's Restaurants and the Cheesecake Factory.

4. **Chain restaurants:** The restaurants in this category do little cooking from scratch, and most, if not all, of the menu is corporate-driven.

As you can imagine, chefs at cutting-edge, upscale casual, and value-driven restaurants encounter more culinary challenges and tasks on a daily basis. In chain restaurants, chefs are more often used in research and the development of new products, an exciting area discussed in the "Interview" section of this chapter and in chapter 10.

## The Feel

### OPENING SCENE

The scene is a dust-filled space with raw walls and exposed ceilings. In the middle of the room is a giant unrefined board covered by crisp white architectural plans. The precise tiny blue lines represent the master plan of doorways, kitchen equipment, seating arrangements, and workflow. A voice from across the table interrupts my thinking. "Everything look in order, Chef?" At first glance, these clean, crisp drawings would awe anyone with their professional appearance and military precision, but as I painfully understand each carefully planned line, I find the hidden mistakes most kitchens take to their graves. This is my first newly designed kitchen, and I plan to recognize those flaws from my years of working these disastrous kitchens. "No," I quickly answer, without hesitation. "I need to walk the space and plan each station as if we were cooking our first meal."

## THE RESTAURANT SCENE

The opening of a restaurant is like the birth of a child. You have nurtured your dreams, shaped your experience, and perfected your cuisine. Your interest in the project includes the dining room décor, color scheme, carpeting patterns, seating arrangements, bar and lounge atmosphere, not to mention the style of china, silver, and glassware. The list of kitchen smallwares grows with every breath of creativity and menu design. The months before were filled with sleepless nights when you mentally prepared menus over and over again. The final cuts were meticulously executed in a peer's kitchen or your own home, with friends and family as taste testers. Many items are classic dishes you've cooked hundreds of times—with a new twist of excitement, of course. It seemed like a lifetime to get to this point—countless hours, immeasurable months, and vaporizing years—to finally hear the words of stature: "The Chef." I'm only 28, yet I've matured in a practice filled with mistakes and successes. I cooked day and night, weekends and holidays, and, usually, on my days off. If I wasn't cooking, I was talking about cooking. Now, today, I have an owner who believes in my spirit, my style, my ability—and, most important, my food. Every inch of the restaurant must be perfect, the concept exact, and the menu inviting and realistic. We need to please, surprise, and satisfy with every bite. We must create an experience and supply a value so when our clients leave we hear the musical words "We'll definitely be back."

# A Day in the Life

8 A.M.: The phone is ringing, and I feel like I'm saying hello, but I'm dreaming. The phone continues to ring. In my half-conscious state, where the mind is willing but the body feels last night's exhausting covers, I grab the phone and say hello, wondering who dared to call this early. Static, and a voice saying, "Yo, Chef, I got the gold." I can't make out who it is or what is being said in my semi-awake state. "Who is this?" I ask. "It's Jimmy from J.J.'s, the winters are in, and you're the first to know." The sleeping chef in me comes to life, wide-eyed and fantasizing about my next creation with these morsels of black magic. "Get me a pound," I eagerly reply. I laugh as I hang up. "Truffles," I sigh, eyes open with anticipation of the day ahead. My husband lies next to me, unaffected by the episode. He is my Sous Chef, equally committed to the passion that drives us in our profession. I'll get up and let him sleep in.

The pitter-patter of little feet comes running across the house. "Mommy, Mommy, are you UP?" It's my five-year-old son, bright-eyed and bushy-tailed, getting ready with the housekeeper to go to school. I make it a point to see him off on the bus every day or let him sleep in with us a few times a month. The life of a restaurant chef isn't conducive to the stereotypical American family schedule. You eat dinner many nights at 9:30 or 10 P.M.; you work nights and weekends, and with luck you grab a slow Sunday off, depending on your restaurant business. You miss many family gatherings unless you can carefully plan with coverage for the Chef, which is difficult. Many restaurants close to give the Chef and crew time off for vacation. The lifestyle is addicting. Your routine, even with children, revolves around your work schedule and allows minimal compromise. So begins another power-filled day. The home is secured, the boy is off to school, and my mind races with newly imagined dishes.

## MORNING PRODUCTION

I arrive at the restaurant at 9:15 A.M.; the buzz of the morning's production is well underway. The scent of veal stock fills my nose with the aftertones of lobster bisque. Just then another scent disturbs these priceless aromas. "What's burning?" I blurt out with fear that a hard-earned sauce or soup has been trashed. "Just a burnt pot of coffee," replies a waiter. In a split second, with a dark roast in my hand, I'm creeping through every inch of mise en place—or meeze, as it's known in the industry—for the lunch menu. I'm called downstairs for every order arriving, to weigh, touch, smell, taste, and return if needed. It's Thursday, and the weekend prep must begin. The setup for lunch is stocked and ready for service to begin. The maître d' stops to review the lunch specials, carefully taking notes and asking questions so he can instruct his well-trained staff.

## AFTERNOON RECAP

Next, a quick recap of yesterday's sales, purchases, and expenses, together with another flavorful java. Price, cost, waste, utilization, and profitability are real words you must live by daily. Indulging in underutilization, excess, and extravagance doesn't work in our occupation. I love the best—but still we need to make money on every dollar earned. We are here for all the above and more. We are magicians in our field, taking simple goods and creating masterpieces. My train of thought is interrupted by the hum of lunch service. The typical business crowd demands value, quality, and atmosphere. In our 70-seat restaurant, we work diligently to please the masses by constructing cool sandwiches, innovative appetizers, assorted organic greens topped with goodies, entrées with sparkle, and desserts to die for. We want guests in the seats, not dust on the tables. The key is to understand what customers want to eat and then figure out how to entice them in a style that pleases you and suits the kitchen's philosophy. Don't ever make the mistake of thinking this is easy!

The turn from lunch to dinner is quick. There is a mellowing silence before the next meal period hits. The kitchen is madly prepping all the final details of a wonderfully planned day. We are anxious and excited, anticipating the rush of 150 reservations already on the books.

## DINNER

Dinner is elegant and trendy yet comfortable, with style and class. The bar will rock with guests dining on a wide range of interesting appetizers that express

The Feel | A Day in the Life | The Reality | Professional Organizations | Interviews | Organization | Education Path Advice

9

the culinary path of their creator. Experience is something a chef can always fall back on, an archive preparation from years past, a specialty from a previous job. Knowledge is the secret that keeps the adrenaline and enthusiasm flowing from day to day, year to year.

The staff prepares a casual and tasty family meal for themselves. All sit and enjoy the last supper before the wave of the evening's reservations. Headwaiter Andy walks around collecting a buck from staff members predicting the evening's covers. One hundred fifty now on the books, a 20-top with a set menu at the same time as the bulk of the reservations. You get used to everyone coming at once. It's important for the staff to keep cool, calm, and collected so the dinner chaos in the back goes unnoticed out front. I review the specials with the waitstaff and do a plate presentation and tasting. The front-of-the-house staff is my link to the customer. They must be educated on my style and philosophy as well as my passion for pleasing.

Just 20 minutes to showtime, you've been cooking and tasting all day. You scan a quick panorama of your kitchen: Production is cooled and put away; cook's mise en place filled, ready, and displaying their magnificent natural colors; the cooks are hanging, telling war stories of other busy nights; sauté pans are neatly stacked like a Doric column only an arm's length away from the stove; refrigerators are stocked with seasoned proteins awaiting the first click of the POS; preparations on the cold station are lined up perfectly like soldiers in battle formation waiting their first command. It's a feeling you must live, a feeling that mixes pride, perfection, obsession, and the need to delight your guests. We strive to be the best we can, and we try damn hard.

## EVENING SERVICE

There is a flow through the evening's slam; we work meticulously to expedite each diner's request. Like a conductor in front of an orchestra, the chef controls the tempo, dictates the beat, and balances the rhythm. The line fills with dupes, table 6 appetizers for 7 people, table 8 pick up entrées for 4, table 15 VVIP special amuse, table 5 rush an extra mushroom flan, table 12 steak black and blue, Bar needs 2 appetizer samplers. When it's on, it's on, sending a cool rush of accomplishment through your entire body. The last of the dinner tickets is completed. The cooks' challenge all evening is to keep their station as organized and neat as possible. This shows their skill level and intellectual preparation. When the last ticket is out, we all high-five each other. "That was awesome," I comment. We all feel pretty high from the great success, but there is still dessert and cleanup. The hum of the dining room is still in full swing with last-minute requests, wine service, and chitchat. I make my way among the tables, greeting, smiling, and making small talk with those guests who are looking for my attention. You want to be available, approachable, sincere, but not annoying to your clientele.

It's 10:25 P.M. I wonder how my son is doing. He's at Grandma's tonight, I quickly remember. I make a mental note that on my day off we will take him to the zoo. What a great kid he is. My thinking is interrupted by the maître d'.

"Great job, Chef! All is well!" "Thanks! Why don't you buy my staff a beer? They deserve it," I reply. I personally walk around and thank everyone for a great evening. Dishwashers first, for they are my backbone; many of them grow into prep and cook positions. Tools are sharpened and put away, ordering finished and called in, cleanup started, prep sheets put in order for tomorrow's production. The cooks are organizing, wrapping, labeling, dating, icing fish, and carefully putting away their preparations. All seems calm; I think it's time to leave. It was a great day, but tomorrow is just around the corner. "Wait, Chef," calls out Craig, my Sous Chef. "You won the pool with 142 exact. Twenty bucks is yours." From halfway down the parking lot I yell back, "Keep it, Cheshu [his nickname]! You deserve it! See you tomorrow."

# The Reality

This segment is probably the most fun, at least for this author—a restaurant chef for about 20 years. The level of creativity and thought applied to your cooking is challenging at every turn and drives the energy of the entire staff. You set the rules regarding style, culture, and philosophy. Of course, you need to make money, so there are plenty of business concerns for which you are also responsible. Even so, the ever-present motivator is to create and serve great cutting-edge food to your guests. You are the artist, the mad chemist behind the dishes that fill your customers with pleasure and force all other matters to the back of your mind. The stove is your workshop. Life is good when it is full and the sting of the heat hits your face.

Naturally, the romance is governed by the fact that you are here to make money, and so is everyone else employed by the restaurant. You must constantly keep in mind that if your art does not sell, the people who are your support and co-conspirators will go hungry. In this lies the challenge of being cutting-edge and meeting a need for a share of the market that ultimately must be substantial enough to support your work. All too often, this little detail is overlooked until it is too late and the ship goes down. The people who come through the door do so for many reasons and have any number of expectations; you as a chef must understand this well enough to hit the mark and drive sales through your heart-felt efforts.

## THE GRIND

You are in by 9 or 10 A.M., depending on the work and the way you organize the staff to perform your prescribed tasks and make ready for lunch service. You probably chat with the lunch sous chef and the cooks and begin to taste their work—all the sauces, marinades, dressings, and other preparations—to make sure they will represent you in a fashion you are happy with. The mise en place must be inspected, because the freshness and good condition of your groceries are paramount. Today you will work on the stove for six or seven hours, stopping

The
Feel | A Day
in the Life | The
Reality | Professional
Organizations | Interviews | Organization | Education Path
Advice | **11**

to review sales reports and the day's purchases, and then glide into the blood pressure–raising rush we call service. There's not much time to waste on rest or breaks. No matter; you probably won't even know when the sun sets. You are happy to be surrounded by the group you know as "the family" and still are not sure why anyone would call this work.

## PHYSICAL ISSUES

Steam, hot oil, and particularly bubbling butter are very hot and will instantly create the pain with which you are all too familiar. Your knives are sharp; you wield them quickly, and even with the years of practice, sometimes a julienne goes astray and you slice a digit or two. Between the rush and the nerves, it is bound to happen over and over, especially in the early years. It is part and parcel of your job, and after a couple of hundred burns, they don't really hurt so much (or that part of your brain refuses to feel them anymore!). This is a really physical job. Your hands are in ice-cold meat for 30 minutes, then over the stove in the blazing heat; next, you yank the 40-pound stockpot off the stove and carry it to the sink, then bend too far over and strain your back. The body-wrenching positions never seem to end. This goes on day after day and year after year, until one day you realize you are not so young and invincible anymore, so you decide that maybe the young cook you are so diligently teaching should help you pick up that pot. Understanding the physical strains and urging your staff to pay attention to them are your responsibility. You probably should have gotten smart sooner, but when passion is the force behind what you do, it is easy to be blind.

## THE BENEFITS FOR YOUR CAREER

The things you do stimulate all the senses in your body; your work brings pleasure to your customers and satisfaction to your staff. You are the person responsible for this business, this place where people come to nourish themselves and be part of your family for a shift. You set the tone for the staff, who work together as brothers and sisters, forgetting the occasional dispute and keeping in mind the sense of success they feel every night after service, when you, the chef, gather them all, have a cold one or a glass of wine, share some food, and thank them for their attention to the night's work. By the time you all say good night, it is late, and tomorrow will be here soon. The ability to do this type of work day in and day out is a testament to your capabilities and talents, which may take you elsewhere some day.

## YOUR LIFE OUTSIDE THE JOB

As in any career, you must build a balance. You have a family outside the restaurant family, and it is important that your management skills be strong enough for you to build a ship that can float without you. It is tough in a restaurant because, to varying degrees, it is built around you, your thoughts, and your abilities. If you spend time creating and teaching the culture while grooming your staff and sous chef, then you will be able to get away with minimal damage. It is ultimately necessary to focus on this aspect of your work, as the time away from the job is nearly as important as the time on it.

# Professional Organizations

Please see Appendix A for more information on each of the following professional organizations for culinary and foodservice professionals working in restaurants.

American Culinary Federation

National Restaurant Association

National Restaurant Association Educational Foundation

Women Chefs and Restaurateurs

# Interviews

### WALDY MALOUF, Chef/Co-owner, Beacon Restaurant

#### Q / Can you tell me a little bit about the history of Beacon?

*A:* We opened the first Beacon five years ago in Manhattan. I was the Director of Operations and Executive Chef at The Rainbow Room with the Baum Emil Group at the time when we decided not to renew our lease for various reasons. Mostly it was just too expensive. Joe Baum and the group came to me and I told them I wanted to open two restaurants, a casual one, and then later on, a more formal one.

The first restaurant, Beacon, would be casual and based on the sort of foods that I enjoyed and had experienced in my travels to places in the Mediterranean and South America. I had this concept of high-quality ingredients, simply prepared yet still sophisticated, and cooked with open-fire grills, wood-burning ovens, and rotisseries. At the time, the market seemed to warrant something more like that, as opposed to a more formal, expensive restaurant. The Baum Group agreed, and I became a partner with them. We opened Beacon to excellent reviews. We've enjoyed the good fortune of success. We just got over the five-year hump and I know it will continue to be successful.

#### Q / I know you're involved in more than Beacon. Tell us what else you do.

*A:* In addition to operating Beacon, I do a lot of other things. I'm in the process of opening a small pizzeria on Sixth Avenue. This is a prototype pizzeria using a wood-burning oven.

*I've written two cookbooks, and I'm in the process of writing a third one. The first cookbook,* Hudson River Cookbook, *was based on my cuisine at the Hudson River Club, where I used products and produce from the Hudson Valley.* High Heat *is about grilling and roasting, the common cooking methods at Beacon Restaurant. It is geared toward the home cook more than the first book. The publishers have asked me to do another grill cookbook, in a dictionary format listed by ingredient. Lastly, I am also working on* Slow Burn, *which is about braising and slow cooking.*

*I have a consulting company called Waldy Malouf Hospitality Concepts. I work on independent consulting jobs for food companies, restaurants, hotel chains, resorts, airlines, grocery stores, and food production. I am consulting for a wine bar/restaurant in Chicago and for Northwest Airlines. I am involved in numerous organizations, Windows of Hope being one of my personal favorites. I'm involved with a number of schools and charities. I am an advocate of giving back and being part of the community. It's something that you do for more than just the networking or the publicity.*

### Q/ What steered you into this business?

*A: When I was 13, I wanted to buy a motorcycle, so I went to work in a pancake house as a dishwasher. I quickly changed positions when I found out I could make $1.00 more an hour if I worked the pancake grills. I liked working the grill.*

*I have always been interested in food. I had a Sicilian grandmother and a Lebanese grandmother and grandfather. My mother is a New England farm girl. I grew up on a farm my family owned in Massachusetts. We later moved to Florida. We kept the farm, and visited every summer. I had a natural affinity with the farm, animals, and the food my family cooked there. I enjoyed cooking, but at the time it was not considered a serious profession.*

*When I was 15, I worked as a dishwasher at a large country club that opened near our home in Florida. The kitchen was run by a French chef and a German chef who had trained in France. They had recently come to the United States, and were looking for an apprentice. They offered me the apprenticeship. It paid more than washing dishes, so I took the job. They took me under their wings, and I enjoyed it. I would work the breakfast shift before school and then I would come back and work an early dinner service after school. I did this all through high school and never thought anything of it. It was a way of earning money. I continued to explore different positions within the kitchen. I enjoyed the camaraderie of the kitchen, the work, and the work ethic.*

*Although I did well in school, I was never the studious student type. When I was about 17, a new clubhouse at the country club opened up, and I was put in charge of about 30 people. It was only open for breakfast, lunch, and dinner on the weekend. It was in this position that I realized I was good with people. I had learned from other people how to supervise, work with people, and get people to do what you needed them to do.*

*Then I graduated high school, and went to college with the idea that I was going to follow in my father's footsteps. I went into pre-law but continued to work in restaurants. I moved to Tallahassee to attend Florida State for my second year. I went to work for another country club that had a classical chef. One night we hosted a dinner for the Confrérie de la Chaîne des Rôtisseurs (a professional association). A representative of the Culinary Institute of America was there drumming up the students. He came into the kitchen and talked us into coming to visit the school. I visited the school out of curiosity. I loved the facility and program so much that I decided to enroll. My parents basically disowned me because I quit college and went into a career that at the time they did not believe was acceptable. I had a strong attraction to culinary arts. Thinking back now, I believe it was the discipline that attracted me.*

**Q/ What did you do after graduation?**

*A: I graduated in 1975 and took a temporary job as a Banquet Chef at the St. Regis Hotel in New York City. Then I worked as the Assistant Saucier at the Four Seasons, and became the Saucier a couple of months later after the Saucier quit. That was one of the greatest jobs I ever had. The Four Seasons was a very professionally run restaurant and kitchen. Everything was prepared from scratch, and menus were changed four times a year and were ingredient-driven. The food and service were serious business. I worked with a high-level professional team of people. It was here that I realized that as much as the cooking was important, so was the front of the house, service, marketing, promotion, and advertising. The food is a major part of a restaurant, but it is still just a part of the restaurant. I also saw that anybody that was very successful, no matter what their position was in the restaurant, always acted like they were owners. They didn't treat it as if it were just a job. It was a lifestyle.*

*After working at the Four Seasons, I was offered a position to open a restaurant in Key West, Florida, as the Executive Chef of a big restaurant at the Pier House Hotel. I put together a kitchen, the menus, and got involved in the design, construction, and marketing. From Florida, I went back to New York to work as the Executive Sous Chef at the newly opened King Cole Room at the St. Regis Hotel.*

**Q/ It's interesting that you haven't said that you ever interviewed for a job. You never had to look for a job. You had a network and connections, so you fell into jobs.**

*A: I've interviewed but not applied for jobs.*

*I stayed at the St. Regis for about a year and then became the Chef of a new restaurant, Christopher's on East 63rd Street. From there, I worked as the morning Sous Chef at Le Côte Basque, one of the oldest, best-known, and finest New York City restaurants. It was an extremely competitive work environment.*

*I took a position as Chef in a small 45-seat restaurant by the United Nations that had no written menu. Basically, the menu changed daily and consisted of four appetizers, main courses, salad, and desserts. We offered a four-course lunch and four-course dinner. The kitchen staff consisted of a*

| The Feel | A Day in the Life | The Reality | Professional Organizations | Interviews | Organization | Education Path Advice |

**15**

*dishwasher and myself. It was truly an interesting experience. I worked there for about a year, basically by myself. It was this solitary time that helped develop my cuisine. There were no menu restrictions other than it had to be good. There was always a fish, bird, and meat on the menu. You did everything yourself including desserts and butchering. There was no one there to help you, so discipline was crucial. I developed my own style and perfected kitchen organization.*

*A year later, I was offered a position at La Crémaillère, a French country restaurant in Banksville, New York. I stayed there for seven years.*

### Q / Did you wish you had done your bachelor's degree?

*A: I still wish today I'd done a bachelor's degree. I have taken business courses at New York University, the New School, and art classes at the Japanese Institute and the Japanese Society. I have educated myself, but I probably would have been able to get into ownership sooner if I had taken the business courses earlier. It has taken me a really long time to get where I am today.*

### Q / What kind of advice would you give young culinarians?

*A: Education is incredibly important, as is your ability to work with other people. As much as the chef is involved in the cooking, he/she also has to acquire the skill of working well with people.*

*Knowledge of history, finance, cuisines, and people from around the world has to be much more today than in years past. I would get a solid education before venturing on my career path. There are so many things that I never thought would be part of my job, from architectural design to colors to music, finances, and loans. I could not have accomplished anything that I have without the cooperation and assistance of hundreds, if not thousands, of people during my career. The ability to lead people along will get people to work with you. Your success is almost guaranteed because you can count on other people to help you develop a great team. You can't do it by yourself. So develop leadership. Those leadership abilities are something that should be taken seriously as part of your training and part of your education.*

### Q / What trends do you see developing?

*A: I think one of the things that didn't exist that much before was how much the public and consumers look to the food industry to influence what people eat and what they cook and eat at home. The reason we have better restaurants in the United States today is mostly because customers want better restaurants and are demanding better.*

*There is a thirst for knowledge about food that didn't exist 20 years ago. I think it is ready to blow across the country. Just look at the Food Network, magazines, the higher profile of American chefs. People want to know where their food is coming from, who is cooking it, and why it is being cooked a certain way. People across the board are taking more interest in what they are eating and where it comes from. And I think that is going to overall have an increased impact on how we are eating 10 years from now.*

**CAMERON MITCHELL, President, Cameron Mitchell Restaurants**

**Q / Tell me about your company, Cameron Mitchell Restaurants.**

*A: We are a $100,000,000-a-year multiconcept, full-service company that operates in six states in the Midwest. We have 10 different concepts, 27 different restaurants, and a full-service, off-premises catering company. We also manage a small restaurant group including four restaurants.*

**Q / How did you get started doing this?**

*A: I was the Director of Operations for a small restaurant company before I started on my own in 1993. I left my operations job in July of 1992. It took 14 months to put the first restaurant together. I was 30 years old at the time when we opened our first one. I started with partners whom I needed to raise capital. Today we have over 180 partners, including 30 operating partners within the company. We raised money on several different occasions throughout the course of the company's history.*

**Q / How did you get into this business?**

*A: I've always said I got fired from my first two jobs, which were mowing lawns and delivering newspapers, and I started washing dishes in high school. It was the only job I could hold. So I stuck with it. I came from a traditional suburban high school here in Columbus, where 93% of the kids went off to college. Both my brothers were doctors, but I wasn't ready to go off to a four-year college. So I lived at home with my mom, and I worked for a large regional casual dining chain based here in Columbus, Ohio. I was with them for about a year when I almost got fired. I was put on suspension for three days and probation for 30 days for being late to work. I couldn't get myself out of bed. I was just working for beer money and was kind of a wayward kid at the time.*

*After my suspension, I came back to work and I was late again. Luckily the manager on duty turned the other cheek. That was a Tuesday and thank goodness he did, because on Friday, I was working a double shift as a line cook during the day and a host at night, and during shift change it was just pandemonium. All of a sudden, time froze, and I had an epiphany. I looked across to the hot line and said, "This is what I want to do with the rest of my life." I went home that night and wrote out my goals. I wanted to go to the Culinary Institute of America, become Executive Chef at 23, General Manager at 24, a Regional Director at 26, Vice President of Operations at 30, and President of a restaurant company at 35. That was it. On Friday, I was just going to work, but the next day I was working for a career. My whole world changed. I went from the bottom of the basement to becoming an Opening Team Cordinator for them until finally I was Kitchen Manager.*

**Q / What triggered that decision?**

*A: I love what I am doing. This is what I'm going to do for the rest of my life. I had one more epiphany that was kind of fundamental as well. If we fast-forward a few years, I graduated from culinary school, came back to Columbus, and worked for a local restaurant company. We had one restaurant* when I joined the company. I started as a Sous Chef opening up our second restaurant. I became Sous Chef for this company and then six months later I was promoted to executive chef, so I became Executive Chef at 23, just like my goal was. And I became General Manager at 24 and then became their Regional Director at 26, and then Operations Director. We had six restaurants then. As I got closer to the center of the organization, I had some fundamental problems with the company.

I decided then to start my own restaurant company, Cameron Mitchell Restaurants. I wanted to build a special company that had a culture and foundation that was second to none, with people that loved to work there. And that is what we are still doing today.

So I started with a legal pad and pen and really no money. I had a little money to live on, but I had no money to start a business. It was a lot of hard work and perseverance and courage to go through those times. I worked for about six months on a restaurant project downtown in Columbus. I was just getting ready to finalize the lease and the plans for starting construction. I raised all the money through partnerships for this project, and the landlord went bankrupt at the last minute. The bank repossessed the building, so everything fell through. I was totally broke. I sent my partners their checks back. I explained to them what happened and sent them my new business plan for a smaller project. We were finally able to get it financed. It was a long 14 months to get the first one open. It's been a struggle, and it's still a struggle today. Our cost of goods is through the roof right now, and profitability is down from last year. When you build a restaurant you expect to pull a profit, but it doesn't always happen like that.

**Q/ Can you talk about the management philosophy of Cameron Mitchell Restaurants?**

**A:** At Cameron Mitchell Restaurants, we believe in our associates. The foundation of our company is based upon integrity—each individual adhering to a code of core values that determines behavior and interactions. Treating one another in accordance with these philosophies establishes strong communication and an enjoyable work environment. Ultimately, we believe we will experience continued success by maintaining this company culture.

We have five questions that we ask ourselves:

1. What do we want to be? An extraordinary restaurant company.

2. Who are we? Great people delivering genuine hospitality.

3. What is our role? To make raving fans of our associates, guests, purveyors, partners, and communities.

4. What is our mission? To continue to thrive, driven by our culture and fiscal responsibilities.

5. What is our goal? To be better today than we were yesterday and better tomorrow than we are today.

Our people are the foundation of this organization. When the company puts its people first, the results are spectacular. The tools we utilize and the theories by which we operate all stem from this belief. Superior service comes from the *heart. We realize that our guests will have a wonderful experience only when our associates are truly happy.*

*The value of an individual is never held higher than the value of the team. For the team to function at its greatest potential, all individuals on the team must work in harmony. It is important that no individual disrupts the positive chemistry of the team.*

*To preserve our great work environment, we hire only upbeat and positive people from a variety of cultures and backgrounds. A positive attitude is an approach to a way of life. It is a conscious choice and the driving force behind exceptional service. Everyone is responsible for fostering an atmosphere that encourages positive attitudes.*

*Work should be fun. We have an exciting work environment that is filled with laughter and smiles. Guaranteed fun equals guaranteed success.*

*Quality is built up front and permeates everything we do. We believe that there is no room for mediocrity. If we are not better than the rest, we become a commodity and are chosen only for price. We measure our quality constantly. What we choose to do, we choose to do best. Our quality is 100% guaranteed. We do it right the first time.*

*We foster open and honest communication. Communication breaks down the barriers to success. We uphold an open-door policy. Feedback creates learning, understanding, and growth. The only bad idea is the one not communicated. We communicate with respect and eloquent language. When we are all informed, we can move forward together.*

*We believe in the creative process. There is art in everything that we do, and art is important because some of the most memorable aspects of service are creative ones. Pride should be taken in even the simplest tasks. Time, people, ideas, artistic beauty, and togetherness all define a creative process, which is inherent to our success and to our organization.*

*We are committed to the growth of all our associates, our company, and our community. We believe that without the growth of our associates, we too will become stagnant. Everything changes around us and we intend to change with it, not to be left behind and forgotten. We are committed to the educational process. We believe that learning should last a lifetime.*

**Q/ What kind of advice would you have for young culinarians?**

*A:* *My recipe for success is to get the best education you can. Work for the best company, have the best attitude. Throw in a little luck and hard work, and you are going to be successful. Also, set your goals and aim high. You can change them along the way, but you have to have a map to get started. Seek advice from your elders. I do that all the time. Go and listen to them. My other advice is to work hard when you are 20 or 30 because you just can't do it in your forties. You've got kids, you've got family, you've got other commitments. Sometimes, you don't have the energy. Finally, have integrity and honesty, and manage from your heart. You treat people the way you want to be treated. I always say integrity takes years to build and days to ruin. No one can take that away from you but yourself.*

**ADVICE FROM A RECENT GRADUATE**

Name:        Alexis Kaufmann
Education:   AOS, Culinary Arts, Culinary Institute of America, 2005

> *Explore any and all directions—you never know what you will enjoy, dislike, or what will interest you until you try.*

# Organization

**Figure 1-1**
Sample Multi-Unit Dining
Organizational Chart

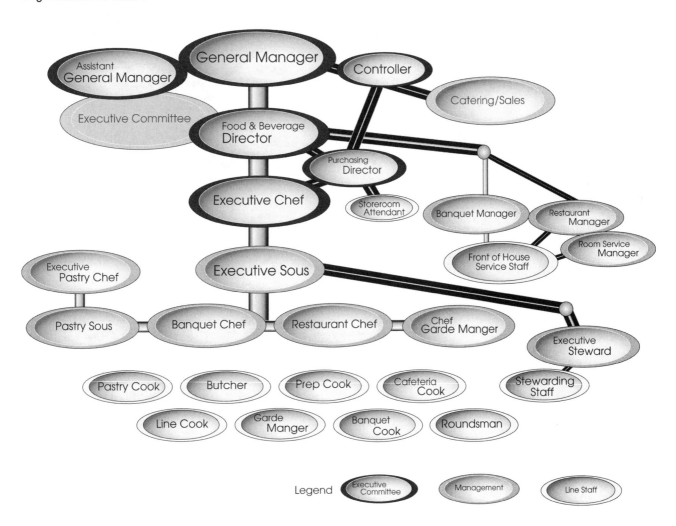

**Figure 1-2**
Sample Cutting-Edge Dining Organizational Chart

# Education Path Advice

**Figure 1-3**
Cutting-Edge
Dining—Education
Path Advice

Height of ball indicates importance.

**Figure 1-4**
Multi-Unit—Education
Path Advice

Height of ball indicates importance.

# EXERCISES

1. At the website of the National Restaurant Association (www.restaurant.org), click on "Careers and Education," then on "Get Advice." Click on "job possibilities" to see a listing of all sorts of restaurant positions. Which position(s) are you most interested in exploring, once you complete culinary school?

2. Using the Career Path Guide (page 2) and Education Path Guides (pages 21–22), map out a possible career and education path for a chef in a cutting-edge restaurant or a corporate chef in a multi-unit restaurant, including:
   - Number of months in each job
   - Each place of employment
   - How and when you would complete your formal education
   - How and when you would complete skills and competencies

3. Find two articles in recent restaurant-oriented publications about culinary trends. Summarize each article in one paragraph.

4. How would you rate the job outlook for culinary jobs in restaurants: excellent, good, fair, or poor? Explain briefly.

5. Would you consider working in restaurants? Why, or why not? If yes, what type of restaurant would you like to work in, and where?

6. Look at the job listings at the National Restaurant Association website (www.restaurant.org). Click on "Careers and Education," then on "Job Bank", click "Find a Job." Print out two job listings that sound exciting to you and that you could interview for while you are in culinary school or once you graduate. Look closely at the Required Skills and Experience section to see if you meet, or will meet, these requirements.

7. Working in restaurants is stressful, the hours are long, and you have to work many nights and weekends when most of the world is off. What would you do to ensure enough personal time to prevent burnout?

8. Go to the library and find five periodicals that relate to restaurants. List the name of each one, and compare and contrast them.

# CHAPTER ② Culinary Careers in Hotels

## Introduction

**HOTELS ARE AS VARIED AS THE MANY** business travelers and families they accommodate. In 2005, almost 48% of lodging customers were people traveling on business. About 52% of lodging customers were traveling for leisure purposes, such as being on vacation. (American Hotel & Lodging Association, 2006) The four basic types of hotels are commercial, resort, residential, and extended-stay.

Most hotels are commercial properties that cater mainly to businesspeople, tourists, and other travelers who need accommodations for a brief stay. Commercial hotels usually are located in cities or suburban areas and operate year-round. Larger properties offer their guests a variety of services, including coffee shops, restaurants, and bars.

Many large hotels have banquet rooms, exhibit halls, and spacious ballrooms to accommodate conventions, business meetings, wedding receptions, and other social gatherings. Conventions and business meetings are major sources of revenue for these hotels.

Resort hotels offer luxurious surroundings and a range of recreational facilities and restaurants. Resorts are located primarily in vacation destinations near mountains, the seashore, or other attractions. Some resort hotels also offer convention and conference facilities to encourage customers to combine business with pleasure.

Residential hotels provide living quarters for permanent and semipermanent residents. They combine the comfort of apartment living with the convenience of hotel services. Many have dining rooms and restaurants that are also open to the general public.

Extended-stay hotels combine features of resorts and residential hotels. The rooms in these facilities usually have fully equipped kitchens.

# The Feel

## MY FIRST DAY

As I stepped into the beautiful oasis covered with tall ficus plants, draping palms, budding begonias, and lavish lawns, I felt at home with a sense of accomplishment. The echo of running water filled the enormous glass-roofed atrium. Multicolored fish swam happily through man-made bridges and tunnels, just as if we were in nature. I stopped to absorb my new surroundings, which I would call "work" for the next six years.

Hotels come in boutique size, elegant and tasteful, or grand, with huge ballrooms and underground roadways traveled by electric golf carts. They are located in every country, in every state, and in every city. Many are tucked in fabulous out-of-the-way places, typically in a comfortable climate. The opportunity to travel and work in these magnificent landmarks is exceptional and well worth the journey.

## RESTAURANTS

Hotels typically have one restaurant serving three meals: breakfast, lunch, and dinner. A portion of the menu is directed by corporate guidelines, and the rest is created by the culinary staff. This arrangement is especially common in the large national-brand chains that offer repeat-guest and frequent-traveler programs. Private or preferred hotels generally allow the Chef total individual creativity. Yet other hotels house multiple specialty restaurants that feature a wide variety of cuisines—traditional steakhouse or American festival, classic French, Asian flair, bistro style, or other concepts representative of the region.

## ROOM SERVICE

All hotels offer in-suite dining, or room service, with simple to extravagant choices that can be ordered around the clock. In the past decade, the emphasis on in-suite dining and an aggressive approach to customer satisfaction have made this department a particularly rewarding employment experience. Daily amenities (baskets of food gifts) have become another great revenue source for most hotels; these are ordered for guests or event clients and are carefully designed by the food and beverage department to reflect the hotel's ambiance or the area's attractions.

## CAFETERIA

Every hotel has a Team Member (employee) Café that feeds hundreds and even thousands per day, breakfast, hot lunch, and dinner. The employee cafeteria may contain full-selection salad bars, cold cut and sandwich stations, assorted beverages, grills with burgers and chicken, or hot dog wagons. The details depend on staff size, budget parameters, and corporate philosophy.

## CATERING

Catering is a hotel operation's biggest revenue source and its backbone. Catering encompasses small VIP all-day business meetings, politically focused lunches, elegant cocktail receptions, and five-course private fund-raisers with rock star entertainment. This is the heartbeat, the income that supports most of the other outlets in food and beverage (F&B). Banquet catering operations are calculated and intense, and they require constant attention. Generally speaking, catering revenues account for about 50 to 70% of gross F&B income.

## BACKGROUND

The hotel atmosphere is corporate, with a team of executive decision makers at each location, as well as a corporate office or owners. The commander in chief is the General Manager (GM), who is poised and confident and exudes years of experience mixed with expertise. The GM is supported by an executive committee that translates the corporate direction, style, and philosophy to line-level staff with the goal of creating 100% guest satisfaction, exceeding all expectations. The culinary staff has a certain status and is perceived as talented, mysterious, dedicated, and professional. With tremendous support from other departments, the kitchen is the hotel's energy source, its heart, and the soul that dictates the rhythm of this gigantic machine.

# A Day in the Life

> ### "It is not the destination . . . but the journey that is the reward."

Now, let's tour this amazing place, a hotel, and discover the opportunities and challenges faced every day by the Hotel Executive Chef. The work is endless, the dedication unparalleled. Skills, experience, and abilities are tested every day. It's 7 AM, and already the hum of the POS (point-of-sale equipment) can be heard as it spits tickets in two locations for the three-meal café. The waitstaff flows from the line to the exit like skaters gliding in a well-rehearsed routine. The breakfast cooks move with planned steps that show their experience and flawless delivery. The chef checks the restaurant and leaves the staff to continue their assignments.

## MENUS

The menus for the day are reviewed along with the prep lists and the food-ordering sheets. I read each daily event to ensure that not a single request is overlooked. The Chef is the core of an operation that touches every department, from the bell service staff that greets the guests and parks their cars to the laundry staff that starches the uniforms to the concierge who recommends the cutting-edge restaurant.

## ROUTINE

As a Chef, the consistency of my routine is paramount, as I must keep track of all the outlets and operations every day of the year. My rounds of the kitchens set the tone for the day. First, a good-morning chat with the banquet cooks and a tour of the coolers and food storage to get a feel for production levels and supplies. A stint in the Garde Manger station to check quality, taste, and timing is next. Then I review the pastry department to check all parties and future events and to address any concerns. The day's events are getting under way, and the timing for the lunch rush is well under control. I stop at stewarding, a thankless department that supports the kitchen's every move. All supplies are shined, clean, and ready for the day's functions. An older man stationed in the dish pit waves shyly and says, "Good day, Chef." I stop to acknowledge his kind words and ask how his two children are doing in school. Respect for all people in all departments is essential to great leadership. Now I'm off to the restaurants to check their setups, daily specials presentations, and quality. A quick cup of coffee in the office, some clearing of paperwork with my secretary, and it will be time to orchestrate the lunch rush. A review of yesterday's sales, payroll, and food expenses is essential to a successful operation, and it's my job to monitor and control them.

## AFTERNOON

Lunch is jamming, yet a well-trained staff can execute flawlessly in the eyes of the customers. The last meal is set before a guest, and my talented staff finishes service while I move on to the afternoon's proceedings. Wednesday is executive meeting day. The top eight people meet to go over financials, concerns, future changes, corporate specs, and important issues that need attention. The executive committee usually consists of the GM; the Directors of Human Resources, Engineering, Sales, and Operations; the Hotel Controller; the Rooms Executive; the Executive Chef; and the F&B Manager. Together we are the decision makers involved in the day-to-day operation as well as future assessments.

Thursdays I meet with the F&B Manager, and every day at 2 PM is the banquet menu reading, where I review all functions with catering and operations. Of course I must guard the financials of the operation, and one of the places where money is spent fastest is the storeroom. The purchasing department is responsible for price bidding, quality control, receiving, and tracking purchase orders. As the Chef, I oversee this process because the potential for excess expenditure is high. Now it's time for a quick meeting with my Sous Chefs to review evening business, staffing, and preparation already under way for the next day.

The
Feel
A Day
in the Life
The
Reality
Professional
Organizations
Interview
Organization
Education Path
Advice
**29**

## THE CLOSE OF ANOTHER DAY

Almost before I can blink, the night's events are in progress, and my well-prepared staff is poetry in motion. We are rewarded daily by excellent customer feedback, countless thank-yous, and low employee turnover. I stop to look at my watch; the time has vanished! It's 8 PM, another 13-hour day has gone by, and I still have last-minute adjustments to attend to. At the peak of the season, the days blend together, and most of my executive-level colleagues work six-day weeks, as I do. Sous Chefs are scheduled five days per week, but some weeks that is impossible. In some hotels, a compensation day (an extra day off) may be granted in those situations.

At last I have got the crew settled, the meals prepared, the ordering for the next few days done, the prep lists established for tomorrow's big events, staff scheduled, and payroll submitted. The management teams are all fully informed of their responsibilities. My tall toque comes off, and I rub my face with both exhaustion and triumph. "Good night, Chef," one of my cooks says on my way out. I shake his hand and tell him, "Good job, James. See you tomorrow." As I walk down the long hallway to the exit, my mind wonders. "That's right!" I think with a rush of warmth. "I am the executive Chef in this magical place!"

# The Reality

You, as Executive Chef, are responsible for the activities of a great number of people and for production 24 hours a day, 7 days a week, 365 days a year. No, that is not your schedule, but it is important that you thoroughly understand how to get your employees to function well whether or not you are present. As every good executive knows, you merely drive the boat and decide the course and speed, as well as how many people it needs to operate efficiently. Management of a large culinary facility is full of pitfalls and variables, and the executive Chef needs top-notch supervisory, organizational, and people skills to make it run smoothly. Once you have accomplished this, the rewards are both personally and professionally satisfying.

Did you notice we haven't mentioned cooking yet? As you venture into upper management, the rule of thumb is to never work until everyone around you is working. Although your days of running your own station or working the line are over, the challenges of nurturing and mentoring your staff, helping them become competent cooks, are endless.

## THE GRIND

The days are filled with corporate meetings, and you must be comfortable and confident taking others' opinions into consideration while intelligently voicing your own thoughts. All corporate ventures require teamwork, and you must be a team player to be successful in the culinary field. Your biggest asset is your ability to gain respect through performance, not intimidation. As you train, you will have the opportunity to work alongside many seasoned specialists; this team will challenge, support, and mentor you, and in return you will do the same for your colleagues. The knowledge you gain in this atmosphere in a single year is priceless and can only be learned on the job.

## PHYSICAL ISSUES

My feet hurt. No, my knees hurt. No, I feel it in my back. Good thing I don't have a headache! These are typical Chef complaints. The hurt comes with the territory. The pain grows with tenure, dedication, and achievement. The complaining is silent, as Chefs understand that the job is physically strenuous. The injuries come because you are constantly exposed to fire, sharp objects, dangerous machinery, blazing-hot metal and steam. You will be wounded, burned, and cut countless times—never too seriously, we hope!—but all Chefs have stories that will quickly help you understand why they develop a grand respect for their tools and surroundings. You, as Chef, will be the chief protector of your employees, not only ensuring that they are in safe surroundings but that their training is thorough enough to protect them. Nothing is more heartbreaking than for one of your cooks to be injured. After all, your staff is your family, even when you are at the top of the ranks.

## THE BENEFITS FOR YOUR CAREER

You will absorb the experience of this job and soon become an integral part of the operation. You will see your effect on the grand scheme of things, and you will continue to grow every day. Your accumulated knowledge of sanitation, systematic production, feeding a well-orchestrated meal to 500 guests all at once, the artful creation of the perfect dish, and the ability to work with a challenging team will be put to the test daily. You must be the same talented professional whether you are tired, mad, sad, or stressed; after all, you are the coach, the general, the leader of the many cooks, dishwashers, stewards, and F&B staff who notice your every move and action. Realize that everything you do is watched and evaluated. Your demeanor, your image, is part of setting the attitude of the workplace.

## YOUR LIFE OUTSIDE THE JOB

Life outside the job is critical to survival inside the job. Have you ever heard of burnout? Too much of one thing can make a Chef a boring person. Whether it is raising a family, having a significant other, or indulging in a hobby, you must do something beyond your kitchen. Stability is key to the success of every great leader. Your time outside this type of profession is limited and somewhat dictated by business demands, which makes being a good planner even more imperative. Not to say you will have no external quality of life, but you must work very hard to strike the right balance. You must be proud of who you are on your time off. Take advantage of the perks of traveling and visiting new places. You have earned them. Downtime revitalization is essential for your job and your health in this domain of unwavering pressures, stresses, multitasking, and deadlines. The trick to this part of your life is balance.

## Professional Organizations

Please see Appendix A for more information on each of the following professional organizations for culinary and foodservice professionals working in hotels.

American Culinary Federation

American Hotel & Lodging Association

Educational Institute of the American Hotel & Lodging Association

## Interview

**JOHN DOHERTY, Executive Chef, Waldorf Astoria Hotel**

**Q / What volume of business do you handle?**

*A: The Waldorf Astoria is a $55 million food sales operations hotel. About 70% of that is banquet-driven. Furthermore, we support two restaurants. Oscar's is a three-meal casual American brasserie which serves anywhere from 800 to 1,200 people a day. It has its own kitchen, Chef, and culinary brigade. The Bull and Bear is a steakhouse that has its own self-contained kitchen as well. Another area of food and beverage is room service that has its own à la carte kitchen for 1,400 guest rooms. We have 130 culinary staff, 7 of which are Chefs, 13 Sous Chefs and then the remainder are Chef de Partie, Commis, Tourant, and about 10 or 12 utility personnel.*

**Q / Is a Chef a member of the hotel's Executive Committee?**

*A: Yes, in this company. That is very important because an Executive Chef today is first and foremost a businessperson who represents the culinary department. You need to learn how to conduct yourself as a professional businessperson. In the eyes of the board of directors of a company and the corporate office, or if it's a small company the owner or General Manager, your role is to keep that business financially sound. So, controlling your expenses, payroll, food cost, and providing food that is going to make the business grow as well as provide a profit is your first and foremost objective. If you can do that, then you will have the freedom with your staff to buy fun china, display pieces, and interesting venues for your buffet. Without those resources, you can't improve your product. So, you have really got to be focused on the bottom line. The old philosophy of "I'm here to be a Chef and I take care of the food, and the Food and Beverage Director takes care of everything else" is long gone. That is a clear road for disaster for any Chef.*

**Q/ When you first decided that you wanted to do this, what road did you take to get to your present position?**

*A: The story started when I was 14 or 15, pumping gas on these long gas lines in the 1970s. It was a freezing cold winter, and I was miserable. My friend said there was a busboy opening where he worked, and I said, "If it's warm, I'm there." So I went and talked to the boss and was hired on the spot. First day at work, they sent me to the kitchen to wash dishes. I didn't care, it was warm. That lasted about two weeks, and then they kept pulling me to help them with the food prep, which I didn't mind. It was fun. Back then it didn't matter what I did, I was making money. Next I got to do cooking. It was a real mom-and-pop restaurant. They did everything from scratch. Only later on in my career did I really appreciate what they did, the pride they had. I noticed right away the pride and the care that they took in preparing food for their customers. It was inspiring, and I really loved it. The owner saw the interest that I took in their business. I noticed right away the pleasure I got in preparing food, taking the care, making people happy, and getting the positive feedback right away. I found the passion in this part of the business.*

*Next I got a job in another restaurant that paid more money. It was a very busy restaurant. This is where I really learned to work hard and meet the challenge of a fast business, yet still take pride and care in everything that I did. My next-door neighbor was a professor at Nassau Community College, and he told me there was a school that you could attend to actually learn to do this for a living. So I looked into it, and I learned about the Culinary Institute of America (CIA) and that was it. I was a junior in high school and that was when my decision was made.*

*At that point I started going to a technical high school in the afternoon. At the tech school, the Chef enrolled me in statewide competitions and doing the hotel show circuit. I really had a good jump start in my career. By the time I got to the CIA, I had already been cooking for probably close to three years and had done food shows and competitions and even won some awards. I was well prepared, which made getting through school easier. I don't remember why or how, but somebody gave me the book about the French Pavilion. That inspired me to only want to work at the greatest places. One of the Chefs at school in my first semester asked me what I wanted to do, and I said, "Someday when I am 30 or 40 years old I hope to be good enough to work at the Waldorf." He said, "You get down there this weekend and you tell the Chef you will do anything. Tell them you will peel onions or sweep the floors, and he'll give you a job." So, that is what I did. I came down to the city, bewildered as a 19-year-old kid can be, and looked up at the Waldorf and went and held my breath to find the Chef. I told the Executive Chef, Arno Schmidt, that I had this externship program and I only wanted to do it there, and I'd do anything, whatever it took. Believe it or not, he gave me a chance at it.*

*So I came here at 19 years old, between my first and second years at school, and worked in all the different kitchens. It was a crazy, crazy place. It was a monster. The amount of covers that we did, we did parties up to 5,000 people, it was a crazy, exciting place but I loved the energy. I loved the potential that it had. Every day the Chef would come out and do a special party for somebody in The Towers. It was very special food, and that is what I wanted to be part of. I did my externship and went back to school. I had it all set up that when I graduated I would come back to work here. Then, I graduated on a Friday and came to work here that Monday. That was 26 years ago.*

*My resume is on the back of my business card, and there is still some room.*

**Q/ What kind of foundation does a Chef need nowadays?**

*A:* *I think if somebody wants to be a Chef, the foundation of that career is cooking. The saddest thing in our business is people who go through culinary school, graduate, and think that they are now qualified to be a Sous Chef. They get out of school and take these jobs as Sous Chefs, and spend a lot of time managing processes. That is what Sous Chefs do besides cook: manage processes, anticipate problems, and work the problems out of the operation. The downfall is they never get enough time to develop their skills and palate. They never learn how to be a great cook. If their passion and their dream are to become a great Chef, if you want to be an authority and an expert in what you do, you have to build your foundation.*

*School is an introduction to concepts. You still have to think and remember what you have learned. It is not instinctive. So for that knowledge to become instinctive, you have to go out and practice it and develop those skills so you don't even think about it. It is the same with your palate—it takes years to develop your palate. The more depth a person gives their education by cooking, the greater they can be. I use the analogy of a skyscraper, and if your education is the foundation, the deeper it goes, the higher you can build the building.*

*A Chef also needs to take advantage of opportunities and make a difference. What are opportunities? Opportunities are where the operation is less than what it should be. It could be an opportunity to fix something, for example. I tell cooks to make a difference. That might mean walking down the hallway and picking up a piece of paper on the floor. Anytime you see the garbage overflowing, you will pick up the phone and call the steward to get it. Anytime you see a light bulb out, you are going to call engineering. Silly little things like that. Why wait for somebody else to do it? Make a difference, and that means taking the initiative. The same thing goes with food. You see an opportunity to make the food better, make a suggestion. Don't come with complaints. I don't like when people come with complaints. Come with an observation and solution and be willing to get involved to correct it. I've got so many things going on, I don't need another problem on my shoulders every day, although those are important things. Come with solutions, be prepared.*

**Q /** **Are schools preparing students well for the workplace?**

*A:* *Yes, they offer many more courses, and the facilities are ten times better than when you and I went there. The one thing that hasn't changed is people. Again, people need to take their careers into their own hands. That's a problem today. Too many people have expectations that they need a mentor or that they are owed something. If you are going to wait for somebody to push you along in your career, you are never going to reach your goal. So set your goals, set your targets, and make a plan for what you've got to do to get where you want to be, because nobody is going to give it to you.*

**Q /** **What do you love most about your job every day? Now it's 26 years. Obviously you are an incredible self-motivator and driving force.**

*A:* *I love coming here every single day. I bring work on the train. I bring it home. I'm obsessed with it. I love the challenge. I love my team. I love the excitement that this place brings. I love the constant need to be creative and to change and develop myself. So we are constantly reestablishing our goals. It's an awesome challenge.*

*I told my Chefs last January that by June I want our customers to eat and ask the waiter, "Is there a new Chef here?" I want dramatic, bold change, and that doesn't come easy. You have to forget what normally works, and you have to try new things and be around people who are willing to take the risk. It's effort. You can do anything. You just have to be willing to work harder. And you have to surround yourself with people who are willing to work harder also.*

**Q /** **What are things that maybe you don't like so much that are challenges on a daily, weekly basis?**

*A:* *In any large operation, it is difficult to make changes because you have to change behavior. The bigger the operation, the more you have to change people's behavior. To create a culture where people buy in and commit to the change is very difficult to do. In a small operation there are fewer people to communicate with, so it's easier to watch what everyone does, to follow up and hold people accountable. In a big operation, that is really tough. Sometimes I wonder how much faster and better I would be able to make this kitchen if it was half the size. Then on the other hand, I don't know if I would find it as stimulating, interesting, and diverse, because what we do here is different every day.*

**Q /** **What do you see for the industry at this point and in the future?**

*A:* *Trends come and go, so I've always been reluctant to jump on any trend. But you look at trends and you try to take something from that trend that you could infuse into your cuisine to make it popular to the mainstream. And that is what we have to do in a larger place like this. So that keeps you fresh and new, but you have to be very careful. American cooking is going to be less about a theme and more about the individual style of the Chef—whatever his or her personal blend is. And I think that is going to continue because the Asian, the Latin, the French, Mediterranean, and American regional influences aren't going away, so you are*

The Feel | A Day in the Life | The Reality | Professional Organizations | Interview | Organization | Education Path Advice

**35**

*going to find some places that do one or the other and then you are going to find a lot of Chefs that combine two, because that works. And the concepts get a little cloudy, but people seem to be willing to deal with that as long as the food is great.*

**Q/ Do you have anything you would want to tell a young, aspiring culinarian?**

*A: Dream about what is going to make you happy and satisfied, and then you have to plan the steps that are going to take you there. Because you are not going to get there unless you have the knowledge and skills to do that job. Life is all about making decisions, thoughtful decisions. Understand that every decision you make has a consequence and potential for greatness, so try to make a decision that is going to get you where you want to be. Along the way, you need to treat people with the same respect that you want to be treated with. Honesty and straightforwardness are valuable in business today. If you are confident, make the right decisions, work hard, and do a good job, you will never ever get sucked into a political situation. Workplace politics is driven by incompetence. It's people covering their butts and trying to blame other people for things that have gone wrong. Make the right decision, give your best effort, do the best you can. You won't always win, so when you don't, it's okay to say here's where we went wrong, here's what we have learned from it, here's what we are going to do different starting tomorrow. Also, be a team player, and by that support everybody around you and treat everybody with respect. Work hard and everything will fall into place. I promise.*

## ADVICE FROM A RECENT GRADUATE

Name:       Gabriel Gomez
Education:  AAS Culinary Arts, Houston Community College, 2006

*Make sure every time you touch food, you put your heart into it.*

# Organization

**Figure 2-1**

Sample Hotel
Organizational
Chart

# Education Path Advice

**Figure 2-2**
Hotels—Education
Path Advice

Height of ball indicates importance.

## EXERCISES

1. Describe some of the types of restaurants within hotels and give examples from properties you have visited. What types of hotels seem to offer more restaurant choices? Which offer fewer or none?

2. At a Disney property, you might find a snack bar, a fine dining restaurant, and a buffet with Disney characters. What challenges would the Chef face in each of these dining options? What is similar in each venue? Why is it important to offer several dining options?

3. To see articles from *Lodging* magazine, go to the following website: www. lodgingmagazine.com. Read an article about current trends and issues in food and beverage.

4. How would you rate the job outlook for culinary jobs in hotels: excellent, good, fair, or poor?

5. Would you consider working in hotels? Why, or why not? If you would, what type of hotel would you like to work in, and where?

6. Look at the job listings at the AHLA website (www.ahla.com). Print one out that sounds exciting to you.

7. Working in hotels can be stressful—you always have guests! What would you do to make sure you got enough personal time to prevent burnout?

8. Using the Career Path Guide (page 2) and Education Path Guide (page 3), map out a possible career and education path for an executive Chef in a hotel, including:

   - Number of months in each job
   - Each place of employment
   - How and when you would complete your formal education
   - How and when you would complete skills and competencies

9. Go to the library and find three periodicals that relate to hotels. List the name of each one, and compare and contrast them.

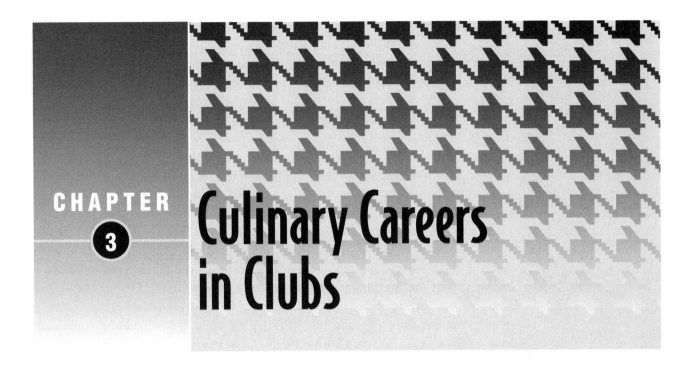

# CHAPTER 3

# Culinary Careers in Clubs

## Introduction

**THE UNITED STATES IS HOME TO THOUSANDS OF CLUBS.** In addition to beautiful country clubs complete with golf course and tennis courts, you can find city clubs, athletic clubs, faculty clubs, yacht clubs, town clubs, and military clubs all across the United States. Whichever type of club you consider, the kitchen and the food it produces are near and dear to the members, as witnessed by the number of places they can eat. As an example, one country club in Atlanta, Georgia, has no fewer than two formal dining rooms, two casual dining rooms, one men's grill, two members' grills, two bars/lounges, three meeting rooms, seven private function rooms, and two snack bars. That's over 20 places to eat! Needless to say, food and beverage revenue is a major part of the club budget, and the Chef is a highly visible presence.

## The Feel

### MY FIRST DAY

I drive to the entrance, which is guarded by two huge 10-foot Corinthian pillars, connected by a trellis-style fence draped with pink baby roses. The plush green rolling lawns, the perfectly manicured flower beds, meet the eye through this magnificent opening. This is grace, elegance, and sophistication, like the grandeur of Tara in *Gone with the Wind*. Images of horse-drawn carriages chauffeured through the entry flash through my mind. At the "club," men were men and women were ladies, and both were lavished with attention from the accommodating staff.

The circular drive ends at an elaborate porte-cochère protecting the guests from inclement weather. The building is massive, with towering French doors that remain open in the summer months, welcoming the cool ocean breeze over the veranda. Balconies with seating line the second floor and overlook the exquisite grounds. The five-story-high weeping willow at the dead center of the rounded drive could tell tales of the diplomats, royalty, stars, and politicians who frequented these grounds in the past 100 years or so.

It's the first decade of the new millennium. The world has changed in many ways, but the rules and traditions of these private country escapes are mostly unaffected. These buildings have encouraged success, built fortunes, and created a model for the social order in America. Huge deals have been made with merely a handshake on the ninth hole of the golf course. Golf is dignified enough to accommodate all ages and both genders. It offers a fair playing field by employing a handicap, symbolizing a culture that has lasted for centuries.

## A Day in the Life

I arrive at the loading dock at 9 AM. A sign at the parking spot right in front has my name stamped on it in bold, unmistakable print. I make my way down the long corridor that mimics the main floor of the clubhouse. This secret lower floor is where the staff moves around the building to the many locations where members await service. The exciting part is that all these service areas connect in the underworld of these glamorous operations. No matter in what kind of food-service operation you work, some form of chaos or sense of urgency pervades the back of the house. This is the world the guests will never see or hear; they just enjoy the result of the staff's dedicated work and effort.

On this Wednesday in June, two weddings are booked for the weekend and three corporate retreats for Friday. Thursday is King Crab Night, not to mention Men's Mixed Grill, Ladies' Bridge Luncheon, Afternoon Tea, and regular à la carte service in the main dining room. My primary mission as Executive Chef is the creation and execution, along with the food committee, of exciting new activities for members to enjoy, both as families and as couples. I am constantly challenged to devise a new outing or surpass one from the previous year. The assignment is to always be conjuring up a new event that excites the membership, maintains their interest in the food and beverage operation, and encourages their attendance.

## PURCHASING

First stop of the day is the purchasing office, to check on all the orders placed. This is the backbone of all food operations, where you can commit expensive mistakes that are quite difficult to identify if you are out of tune with your vendors and pricing. For instance, if an inferior or overpriced product comes into the establishment and no one catches it immediately, nine times out of ten it is used or prepped before the problem is discovered. I work to be proactive in purchasing early in the day so I am well prepared and in harmony with product pricing, specs, and availability. This is basic good practice.

Once I review the next five days' needs, it's off to the main kitchen, making a quick stop in the laundry room to pick up clean and starched whites, share a fast story with the workers about a movie or TV show I saw, and remind them today is Texas BBQ in the staff café. Never forget your support staff, whether they work directly for you or not. A few minutes of friendly conversation is appreciated and shows your respect and consideration for all.

## STAFF CAFÉ

Down the long hallway, freshly painted for the 200th time, I make my way to the next inspection point, "the staff café." This sounds like an unimportant checkpoint, but this foundation area can become a big thorn in the side if not set up with the same standards and professionalism as your kitchen. On a busy day in midseason, the staff café serves 60 to 70 people per meal period, usually including the General Manager, the golf and tennis pros, the office staff, the grounds crew, and all the other staff members necessary to run the facility. Some of these people eat here for three meals a day, and it's imperative that they see you as a great and caring Chef. They rarely eat your food from the menu, so this is their only indication of what and how you cook. Club members often ask staff about the food and the Chefs, so doing a great job in your café is in your best interest.

## THE KITCHEN

Onward to the hub of the vessel, where production is well under way. Stocks simmer happily in 60-gallon tanks, and the Swiss braising kettle, humming with gas flames, is filled with a rich beef stew overflowing with colorful blanched vegetables. The smell of buttery cottage dill rolls, a recipe from my early years of culinary school, permeates the kitchen and hallway. Fifteen years after entering a professional kitchen for work, I still feel a warm rush of enthusiasm and a sense of thrilled expectation for the day ahead. How lucky am I to come to work every day, eat what I want, make what I like, and do what I love. Joe, my morning Sous Chef, gives me a tour of the hot and cold line, reviewing production levels, prep list, lunch setup, and, of course, the specials for the grill and main dining room. There are already 100 covers for lunch in the main dining room and 50 in the grill. As the Chef, I must have my nose in all the staff's business, ensuring that their work expresses my style and my philosophy. All the food they make must meet my standards of excellence.

## THE MIXED GRILL

Known in these social havens as the Men's Mixed Grill, this venue is masculine in design, with huge leather armchairs, giant coffee tables, and Baccarat crystal ashtrays. The bar, of handcrafted mahogany, curves into the matching colored walls like the bow of a grand sailing ship. The room is punctuated with private alcoves lined with thick benches and tobacco-colored leather couches. A row of imported and domestic brown liquors features the Maker's Mark Manhattan, the neat Basil Hayden bourbon, and the savory 30-year Dalmorie, a priceless single-malt scotch. The glassware is strong and sophisticated, as are the china and dinnerware.

The menu is that of a classic New York steakhouse—straightforward, no frills, large portions, and simple yet tasteful selections. Menu classics include colossal shrimp cocktail, oysters Rockefeller, clams casino, the daily selection of shucked clams and oysters, iceberg and marinated tomato salad, and, of course, a Chef's pâté du jour. T-bones, 20-ounce porterhouses, New York strip steak, and thick filet mignon are all served with our home-brewed steak sauce. In addition, we serve grilled 10-ounce salmon and swordfish, sautéed Dover sole, and the unforgettable—2 pound lobster (market priced). Desserts range from the all-time favorite, crème caramel, to traditional New York–style cheesecake, with a variety of fresh fruit toppings, to the sinful eight-layer chocolate cake. Done well, this kind of menu is still one of the most enjoyable and unpretentious, and it fits perfectly into this equation. Picture a packed room full of men, business deals on the table, and a no-nonsense, no-fuss, get-to-the-point dinner. Lives have been changed, decisions taken, disasters avoided, and industry developed over a warm snifter of Hines Rare and Delicate and a fine Cohiba cigar.

The grill is open for an intense business lunch or a midafternoon snack after a round of 18. Hearty sandwiches, inviting salads, and approachable entrées like veal piccata and calf's liver with bacon and onions are the choices in this bit of the Old World in the midst of the modern-day United States. Grills can be found everywhere, from the big-city alumni clubs to the faraway hidden oases tucked deep in the 'burbs.

## EVENTS

The plush ambiance, picture-perfect locations, and grand ballrooms make these venues extremely desirable for special occasions. Everything for social gatherings is already in place at clubs, which now produce affairs like a boutique hotel. Typically, a banquet manager or maître d' handles the booking of the party as well as the scheduling, planning, details, and execution. The back of the house employs a banquet event order like that used by all catering facilities, preparing large events as it supports the other food outlets; this takes careful timing and

The Feel | A Day in the Life | The Reality | Professional Organizations | Interview | Organization | Education Path Advice

**43**

attention. Catering in all food and beverage venues is the gravy of the financial picture. In a club, producing catered events augments membership dues, helps support renovations, and keeps the NOI (net operating income) out of the red.

## HALFWAY HOUSE AND SNACK BAR

"2 dogs, 1 chick wrap, 2 Bobs, no make that 3 Bobs." This is in the language typically spoken by the staff at the summer concessions at most suburban clubs. Primitively laid-out makeshift kitchens have been remodeled to replicate cool sidewalk cafés overlooking Olympic pools filled with hundreds of children during the summer months. Just sign the chit. That is the billing system used to charge the members as droves of kids feast daily on typical snack bar foods. Both poolside snack bars and halfway houses (located halfway through a round of golf) have upscale menus as well. Although the foot-long dog with all the toppings and the double bacon cheeseburger still rule, more creative and lighter alternatives have become standard items. A baby pizza oven cranks out individually designed pies. A panini grill adds a twist to the upscale sandwich selection. Sensational salads with a variety of sensible toppings make an attractive alternative on a sweltering hot summer day. The staff comprises mostly temporary summer employees, but many of them have worked through high school and well into their college years, which is a big help at such a seasonal facility. Many teachers and others in professions that take the summer months off work as responsible and competent supervisors of these very busy outlets.

## THE END TO ANOTHER DAY

As you can see, being the Chef at a prestigious club is both a wonderful experience and a great career. These jobs are among the most desirable in the industry today. Club members care and are knowledgeable about high-quality cuisine, to which they have grown accustomed through their world travel. They crave and desire foods like those they eat at named metropolitan restaurants around the country. With transformation of menu items such as roasted duck with sauce à l'orange into trendy selections featuring flavored oils, light vegetable broths, and reduction sauces, prepared with delicate techniques of cooking and presented with exquisite taste and presentation, has led to new dining room standards

Classics never change; they merely get updated with a new twist and feel. Being a Chef allows you to be more creative, better versed, and more informed than our clientele—because we have to.

It's a beautiful Sunday evening in June. The sun is setting over the 18th hole, shedding a gleaming golden blanket of light with a trail of fire-engine red. It reminds me of the glistening open-fire broiler I stood in front of for the past four hours. The tour of this culinary haven has come to an end, and it's time for this Chef to head for home. It was another crazy, busy day, but the satisfaction of coordinating all the events and executing superb food is an unexplainable feeling worth all the hard work and effort.

# The Reality

The Chef position at a club is perhaps one of the busiest jobs at some points and one of the slowest at others. Nobody golfs with snow on the ground or in 102° heat, but you will be up to your ears in work when the season is on and members swarm the grounds. This is particularly true because in the slow season you lay off a lot of staff or lose them due to labor cuts. Yes, probably every year just prior to the busy season, you will need to hire a good deal of new staff, and then you will train them and lose them at the end of the season. This is an annual challenge; consequently, developing a solid training regimen and clear-cut guidelines is essential. Did I mention that most of these transient staff members will have no intention of being cooks? Young people on summer break who need a job will support much of your culinary operation. This is not necessarily a bad thing; it just increases the depth of the challenge at hand and reinforces your need to be a great communicator, manager, and organizer.

Next, you will find that some clients consider themselves owners or stockholders, and they will not hesitate to communicate that attitude in many ways. This situation is manageable, but it requires a personality that is at the least politically sensitive. You need not have a subservient nature, but you do need to enjoy interaction with people and to be capable of dealing with demanding customers in a calm and confident fashion. These customers will think nothing of devising a menu item and telling you how it should be prepared for them, each time they decide to dine with you. You are a host and the provider of a sublimely hospitable restaurant; particular requests are your specialty. Once mastered, this is a skill of which any Chef can be proud, as it indicates that you have an endless repertoire and thoroughly understand the true nature of our business. Further, once you have won over your membership, they will support you in every endeavor, whether rebuilding the kitchen or buying a home. Remember, club members are often influential businesspeople and society moguls who have the power and knowledge to get almost anything done.

## THE GRIND

At times you will wonder if you can see each outlet every day, and will not be sure how you are going to monitor every activity. Most of the properties are vast, and you will need your golf cart to travel. Some clubs have outlets four or five miles away from the main property. Your job will keep you busy, and you will most certainly not be bored in season. However, once the season ends, at some point you will find yourself standing almost alone in the cavernous back of the house, which was only weeks ago filled with the buzz of your fully charged brigade, dealing with the frenzy of business. Your staff is now almost all laid off or otherwise put on hold. Many will return next year, but a reasonable proportion will have moved on or no longer need the job.

With that said, the other side of the grind is the part of the club that grows old and does not change—that is, the old favorites on your menu. Some menu items are standards (they don't call it a club sandwich for nothing), and you should be understanding enough to embrace these things as if they were your own creations, for your first rite of passage will be to execute them as the past 20 Chefs have. Do not under any circumstances try to change these unless the food committee puts it in writing and sends a letter to the membership, or you will feel their wrath! On the other hand, you can be creative with all the functions and special menu nights, as well as the seasonal changes for your dining rooms. Don't worry—there is almost always room to express your culinarian self!

In addition, as we mentioned above, many private clubs are now opening their doors wider than ever as catering venues to nonmembers, acknowledging that this is a great way to strengthen the club financially without assessing the membership. So, as you may have guessed, this will help keep the place busy in the off-season and mean you can retain some ancillary staff. The catering aspect of club work is a rising trend and a sensible one that is expected to continue.

## PHYSICAL ISSUES

The pitfalls of the culinary world exist in clubs too: the cuts, the burns, the bruises, the long hours that make the feet burn, and the ever-present danger of lower back pain. However, as pointed out already, clubs offer a relatively generous respite as well as a good amount of slow time in which to recoup, regroup, and heal. The expression "no pain, no gain" is appropriate in the club world as well.

Don't forget to watch out for your employees! You will have a lot of inexperienced staff moving around. Preserve their safety through good training and careful supervision.

## THE BENEFITS FOR YOUR CAREER

Once you have established yourself at the club, you will enjoy considerable access to successful business experts. The people who generally frequent country clubs are leaders and learned individuals who will support your endeavors and, in many instances, help you to a better quality of life. In short, never underestimate what they can do for you. As well, you will most likely become involved in the associations that govern the management and Chefs of America's clubs. These associations are filled with some of the best in our industry, and you can gain a great deal of knowledge from their networks.

## YOUR LIFE OUTSIDE THE JOB

You will probably get at least a month off during the slow season—the payoff for the grueling pace you endure in the high season. Club life does offer enough downtime to allow you a reasonable quality of life with your family.

## Professional Organizations

Please see Appendix A for more information on each of the following professional organizations for culinary and foodservice professionals working in clubs.

American Culinary Federation

Club Managers Association of America

## Interview

**EDWARD G. LEONARD, CMC, AAC, Executive Chef at Westchester Country Club, President of the American Culinary Federation, Manager of the 2004 and 2008 Culinary Olympics Team**

**Q / In your job as Executive Chef at the Westchester Country Club, whom do you report to?**

**A:** *I report directly to the General Manager of the club, and informally to the Executive Director. We have a staff of 58 who work in culinary, stewarding, food and beverage operations.*

**Q / What is a typical day at work?**

**A:** *I work anywhere from 9- to 14-hour days. My day starts off by making all the rounds to the kitchens, talking to the Chefs to see what is going on for the day. There are meetings, such as F&B meetings or staff meetings, depending on the day of the week. Meetings might include reviews of the business plans, budgets and forecasts, basically communicating overall production of the staffs and understanding what every department is faced with.*

**Q / I know your club has a large membership. What kind of secondary management do you have in the kitchen, below your position?**

**A:** *The Executive Sous Chef, Banquet Chef, Banquet Sous, Restaurant Chef, Restaurant Sous, Pastry Chef, and Beach Club Chef all report to me and are all in different places, so it keeps me traveling around the club grounds.*

**Q / What is the number-one way that you get your message across to the line employees?**

**A:** *There was a lot of challenge in the beginning. It really took a lot of communication and constant reinforcement. The culture, the philosophy, what we call culinary pride is visible all over the kitchen. In the beginning it was not so much informing supervisors but carrying the message to all and having staff meetings on all levels.*

*Once objectives, policies, and standards are in place, they are communicated to the group of supervising Chefs. The menu-changing cycles are difficult*

| The Feel | A Day in the Life | The Reality | Professional Organizations | Interview | Organization | Education Path Advice |

**47**

*because the problem is that you have to really rely on your supervising people to carry the message to the staff and constantly reinforce the standards you have tried to develop. They are going to have to go through the dishes with all the cooks, constantly reiterate the instructions, and ensure that they are carried out each time that dish is produced. It requires a lot of constant interaction with all levels because it's always important to ensure that everyone still understands the mission and the goals.*

### Q / How many hours do you work every week?

*A: It varies, but probably minimum is 55. It can go as high as 70+. There are opportunities to be successful in this field, but none come without hard work. The hours you need to put in to pursue your goals, and keep it at a superior level, is a lot. There is just no shortcut, there's no easy road to make lots of money and get a lot of titles. There is a discipline of learning the many aspects of the business and working your way up. Find great Chefs to work with, learn our craft, take time to learn how to become businesspeople, and all the while try to progress to that next level. It's a real climb that needs a five- to ten-year game plan of where you are going to go and how you want to get there. It's time and effort that are well spent.*

### Q / What was the career track to your position?

*A: It's been a very diverse, odd career track in terms of being a Chef. I've spent a lot of time in the corporate world and a lot of effort learning ancillary skills. From a young age, I worked my way up. I worked in kitchens in restaurants, country clubs, and hotels for about six years. Then I went into operations, food and beverage management, and sales and marketing, mostly to earn more money and learn new things. I didn't realize at that time how much it really helped me. Even in this job, though it's a culinary position, it is a real management position, and all my experiences have helped me greatly in driving what I have at the club. Understanding the importance of mission statements, philosophy, culture, manuals, numbers, operations, marketing, all these disciplines and attributes have helped me. The more diverse your skills are, the more successful you will be in whatever endeavor you take on.*

### Q / What was it that made you decide to pursue a culinary career?

*A: I think the interest in food, and of course my family. My grandmother, who I was close to, cooked all the time. She was from Italy, and my family loved to eat and enjoy food; it was just so much a focus of life. Everyone was present at Sunday dinners at 2 P.M. I really just always had the interest in me. I also liked art, music, and sports. But I guess when I wanted to make a living, it came down to food and cooking, because those were the things I liked.*

### Q / What would you have done differently in your career?

*A: That is a hard question. I've always been one to move forward. I am very fortunate that I'm just extremely self-motivated, always pushing for excellence. I've always been my harshest critic. I think if you get promoted too early in your career, especially in a kitchen, you may wind up with no one to mentor you.*

*Early in your career, you just need to spend as much time as possible learning, because in our craft there is always something new, something different, or something we have yet to master. Even today, at 44, with whatever people say I've accomplished, there is not a week that goes by when I don't pick something up from somebody. When you judge a competition, you learn something, such as a new dish, or something that is not so good. But it's constant learning. You share with colleagues, you network.*

*But going back, I don't regret anything. There were times when I was going into management, for example, and maybe I didn't understand how a skill, such as public speaking, would pay off for a Chef in the end. However, today I'm very thankful that it actually did happen, and all of those experiences were really valuable for my career.*

**Q / Tell me about your philosophy of management and leadership.**

*A: I've seen lots of good managers. But managers do not necessarily make leaders. It's very important to create a culture, to let people know why they are there, to get people to buy into it. Leadership is that ability to take people and bring them to places that they would never have been without your influence, your assistance, and your help. Leadership means having presence, being able to walk into a room with charisma being evident. I think leaders are people that excel above that management level. When you decide to lead the team, you have to lead them in the direction of good for all, and that direction has to be the mission that the organization needs to achieve. Giving opportunities is another aspect of great management. That is a heavy responsibility, because you are the one who is there to generate success.*

*It is also important to realize that everyone has a life and should be able to live it. You must give people the opportunity to achieve a balance of life. There is something in your employees' lives besides work. That is something that has taken me time to learn, and something that I don't always do for myself. But I definitely do with my staff, because if they are happy at what they are doing outside of work, then they are happy when they are in the kitchen working.*

*Excellence is not something staff can achieve all the time. But I think it is up to you as a leader to instill the challenge of achieving excellence in your staff. If you don't try to exceed customers' expectations, keep raising the bar of what you are doing, then you can never improve yourself or the organization. So, I created the culture of "Yesterday's meal was good, but how are we going to make today's meal better?" What else can we do to take something that everybody enjoyed and make it just that much better or special and create an experience for customers? Everybody looks for a wow factor today, whether it be through your food, ambiance, service, or a combination of them. I think that is always what keeps people coming back: creating an exceptional experience. Most importantly, though, it is generated from the work of the team you have formed, and from basically making sure everybody is on that same page and working toward a common goal. If you have done your job as a leader, then everybody buys into it and believes in it.*

## Q / What do you like most about your job?

*A:* It would be that I have the freedom and the respect from management which allows me the ability to do anything that I see fit, from a great hamburger to classical cuisine, from Italian to buffet work to ice carvings. I like the diversity. I like the challenge. I like the multitasking. I love the variety of challenges. If I went into a job that was the same every day, it would be hard for me to cope. I like the ability to go in every week or every day and not know what's going to happen. That just really keeps energizing me. There are some events you always have, like the holidays. I'm challenged to make the holiday events better than we did last year.

One of the best things is that, at this stage of my career, I am fortunate to get to work with a lot of students. I think there comes a point in your career, regardless of what you have done personally, when you need to give back. You're successful; your business has been good to you; you have an obligation to give back. I am lucky to be shaping young people and encouraging them. It is also a great pleasure to be able to guide young cooks to the type of success where you can stand back and let them have opportunity to shine in the light of it. I think what I like most is that I have the ability to affect a lot of people right now and to give people opportunities that I unfortunately didn't have when I was coming up in this business.

## Q / What do you like the least?

*A:* As much as I love what I do, it does consume me at times. That does take time away from things that are important. My hobbies tend to be my craft, so there are things that I wanted to do throughout my life that I still quite haven't found the time to do. There is time with my family that I miss when I get home at night late and they have had dinner already. That is a tough thing to come to terms with sometimes. It's just the sacrifice that you make in this business. When everybody is enjoying themselves, such as during the holidays, you are there providing for people's enjoyment. So you don't tend to have that joy as much, and that is a sacrifice that you make. That's tough, but I do sit back now and take time for myself a little more often.

## Q / What advice would you give a student who is reading this book?

*A:* Although many students are attracted to fine dining, this segment may be less profitable and the hardest to work in. Other opportunities in the field are tremendous, from research and development to supermarkets to corporate environments. I hope students will sit back, read this book, take the advice, and say, "Wow, I need to learn. I need to have passion. I need to really love this business." This is not the industry where students can go because they don't want to study or learn a lot of facts and details. That's wrong. You need a great education and a love for food to be successful. It's a myth that you don't need to go to school in the restaurant industry. I think, hopefully, they would come out of it saying, "I can be successful and have a life." But if you really want to be super-successful and be the Master Chef or the star Restaurant Chef, that is a huge commitment, and you're going to have to sacrifice. It is going to take a lot out of you, and maybe that

*balance won't be there for a period of time. But for most students who aspire per-
haps to be Executive Chef one day, they will have to study hard and work hard.*

## ADVICE FROM A RECENT GRADUATE

Name:        Robert Geosits

Education:   Culinary Arts, Sullivan University, 2007

> **Read as many culinary-based books as possible and keep an
> open mind.**

## Organization

**Figure 3-1**
Sample Club
Organizational Chart

# Education Path Advice

**Figure 3-2**

Clubs—Education Path Advice

Height of ball indicates importance.

## EXERCISES

1. Investigate two clubs in your local area. What types of clubs are they? What types of clientele do they serve? Clubs that offer on-premises catering often have websites. What kinds of catering does each club do?

2. Compare and contrast the job duties of an Executive Chef in a restaurant with those of an Executive Chef in a country club.

3. Why do you need to have a lot of experience to be an Executive Chef in a club?

4. Would you like to work as a Chef in a club? Which activities would you enjoy? What parts of the job might be difficult or not as much fun for you?

5. Visit the website of Club Managers Association of America (CMAA) at www.cmaa.org. Click on Membership and then click on Club Management Magazine. On the homepage for their magazine, click on Dining Room to obtain a list of articles. Read an article from the list and summarize it in one paragraph.

6. Using CMAA's website, list ten benefits of association membership.

7. Using the Career Path Guide (page 2) and Education Path Guide (page 51), map out a possible career and education path for an executive Chef in a club, including:

   • Number of months in each job
   • Each place of employment
   • How and when you would complete your formal education
   • How and when you would complete skills and competencies

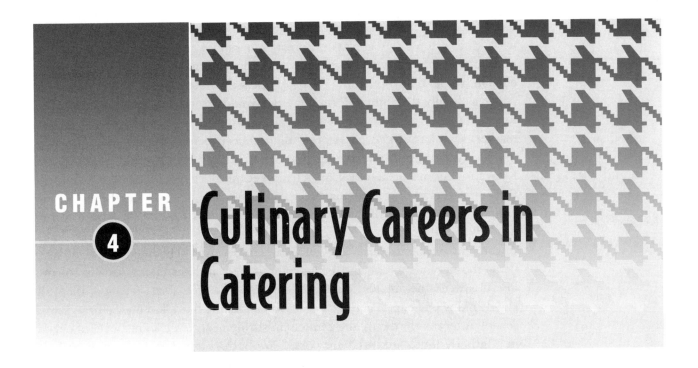

# CHAPTER 4

# Culinary Careers in Catering

## Introduction

CATERING TAKES MANY FORMS, from fancy box lunches at conferences to weddings in the country and corporate affairs in the city. Many catering events are in locations you might never see were you not in this business: country estates, corporate headquarters, galleries, openings of huge department stores and malls, institutions like Ellis Island and Radio City Music Hall. You can cater to people in the media, fashion, business, and the arts, so you can say that many of your events are star-studded. There are many reasons why catering is such an exciting and rewarding career. Every event has its own identity, from the location design to the brilliant tabletop composition to the menus. This chapter is full of details about working in this wide-ranging segment.

## The Feel

### THE JOURNEY OF A CULINARIAN

It was 1985, but I remember my first day at this remarkable catering company like I remember watching the rising of my first soufflé. The excitement went through my bones like a jolt of lightning, yet I had trepidation and apprehension. This was my first full-time position out of culinary school. Weeks before my first day, I thought about what my job would entail, who was on the culinary team, what kind of food they cooked. My restlessness and enthusiasm sent me to my new home several times late in the evening just to make sure I knew the most efficient route. Can you believe they say Chefs are obsessive!

It was finally the night before. I anxiously laid out my freshly pressed whites, razor-sharp knives, polished black clogs, a tiny notepad, and a new Sharpie marker. My plan was to get a great night's sleep before the big day. The only problem was that my eyes would not close and my brain would not shut off. I stayed focused on the clock, which flipped numbers like laps at a racetrack. Of course, the second I managed to get to sleep, the buzz of the alarm had me jumping out of bed. I dashed to the shower, got dressed, and off I went to my new adventure.

The only problem was, it was 6:15 A.M. and I wasn't scheduled until 7:30 A.M. I was too nervous to eat, but I knew about a local diner ahead because of my dry runs weeks before, so I pulled in for a cup of coffee. I sat staring at the wall and drinking one cup of coffee after another, looking like a kid on the first day of school. After my third cup, the gentleman sitting next to me leaned over and said in a fatherly tone, "First day of work?" We both laughed as I commented how noticeable I must look. We made small talk for about 20 minutes, which calmed my nerves and helped pass the time. The gentleman excused himself, as it was time for him to go to work, thanked me for a charming morning, gave me his card, and said to give him a call if I ever needed any paternal advice. I thanked him, and he was out the door. I glanced down at the card—and my knees nearly gave way. The gentleman was the owner of the catering company I was going to go work for! "What are the chances of that?" I thought.

When I arrived at the facility, the receptionist asked me to take a seat. I tried to stay focused and not appear too nervous, when someone tapped on my left shoulder and nearly sent me through the roof. I turned to find the man I had just shared coffee with. We both laughed again as he took my hand and said, "Let me take you around and introduce you to everyone." I knew from that day forward, this was to be the best job of my life.

## THE KITCHEN

I had heard the word commissary before. Examples include military venues that produce large quantities for multiple units, and the massive central kitchens that manufacture products for large chain restaurants. My first days at this catering facility started in the commissary—which, although an interesting locale, is not a glamorous one. The facility was spacious compared to others where I had worked, and the accommodations were impressive. The kitchen was designed expressly for off-premises catering. It held 15 workstations, each 8 feet long, that could accommodate 30+ Chefs and cooks prepping. It had separate walk-ins for raw products and huge walk-ins for storing prepped party menus ready for transport. It featured a cooking battery of convection ovens, a Swiss brazier, steam kettles, and chargrills. It was impeccably organized, professionally ruled, and quality-driven. On the main wall at the entrance to the kitchen were huge corkboards covered with in-depth prep lists, event locations, and staff schedules. On the opposite side was a row of clipboards, one for each day of the week, containing the event orders for the parties. These were the bible for each event, carefully detailed and precise in every aspect: time, date, location, directions, staffing, equipment, pack sheets, menus, etc. This system ensured that nothing was forgotten, because in catering, once you leave the kitchen, you either do without or run to the local store, which gets old rather quickly.

## A Day in the Life

I vividly remember my first off-premises event, just several days after I started working. It took place at the Metropolitan Museum of Art in New York City. I had visited this museum several times to view the exquisite exhibits, but never did I imagine the experience of producing a catered event in such an interesting venue. I was in awe every moment—from unloading the truck in the museum's loading dock to moving our equipment through the gallows of the back hallways to heavily guarded rooms. We set up makeshift kitchens in hallways adjacent to the Great Hall, the Temple of Dendur, and the American Wing, whose walls were hung with priceless artwork. We were forbidden to go from one area to another without an armed museum guard as an escort. I knew, even with my limited experience, that catering a party of 800 is challenging no matter the location; executing this job seemed as if it would require a miracle. But our work appeared to be flawless, and I was incredibly proud of my job and the company I worked for. I remember how impressed I was with the camaraderie of the staff members and how they made sure I felt comfortable in my assignments.

I sensed a standard of excellence the second I walked in the door of the commissary. The kitchen staff was dressed in standard checks with hats and aprons, starched and professional, just like at a culinary school. The shelves were stacked with books of standardized recipes, food specifications, and photos of all the menu presentations, past and present. My first-day experience, although exhilarating, was overwhelming. "How will I keep up with all these people?" I thought to myself. "They are all so knowledgeable, confident, and focused." My fears stayed a secret, and I remembered the words of my first culinary teacher: "Fake it till you make it; never let them see you sweat; be busy; be polite and be quiet, and you'll get it."

I remember the sales and administrative offices on the first floor and the kitchen on the second, which created the cohesive work team that is imperative to the success of any operation. Every day for lunch the staff had a family meal. A designated cook, usually a first-level cook or supervisor, prepared the meal of the day for about 45 people, depending on how busy we were. The cooks presented their goods with pride and accomplishment while the owner of the company said encouraging words about the fabulous job all were participating in. Then we broke bread together, like a family at Sunday-night dinner. It was a great way to pull a team together and create a powerful workforce committed to giving 110%.

The days seemed to fly by as my responsibilities became more challenging. It was apparent that I was passing many staff members who had been there long before me. I was a hard worker and left no one to fall. When I finished my tasks, I helped others finish. I voiced my opinions without sounding conceited or obnoxious. If something needed fixing, I made sure the right people knew about it.

## THE EVENT

Many catered events are in locations you might never see were you not in this business: country estates, corporate headquarters, galleries, openings of huge department stores and malls, institutions like Ellis Island and Radio City Music Hall. You cater to people in the media, fashion, business, and the arts, so you can say that many of your events are star-studded. There are many reasons why catering is such an exciting and rewarding career. Every event has its own identity, from the location design to the brilliant tabletop composition to the menus.

It's a Thursday morning in midseason. The commissary is in full swing, preparing the 18 events scheduled over the next two days. Your workday as a manager depends on the amount and kind of business booked. Typically, when business is strong, the Chefs work 60+ hours per week; in the slower seasons they may work 40 to 45 hours and perhaps have a four-day work schedule.

An event is a three-part process: prepping the raw materials and cooking as much as possible before the event; loading and travel; and final reheat and cooking at the event site. A lot can be prepped ahead without sacrificing quality: mixing composition salads (minus the greens, of course), blanching vegetables, creating the three-potato and apple flan to accompany the sliced filet. The pastry Chefs prepare their goods in advance, short of heating or scooping and, of course, plating. They might have created delicate friandises for each table and dessert consisting of a rich warm chocolate torte with white chocolate sorbet and macadamia tuile. The main courses are usually roasted and cooked at the event site for the best quality possible. As the Chef, you need to plan carefully so your event is technically flawless without sacrificing quality.

The third part of the event is the final execution of the menu onsite. All the cooking methods must be coordinated so everything is ready at the same time.

We are a high-volume caterer and take enormous pride in making everything from scratch. Our quality is the direct result of our ingredients, prep work, production, transport, and final event execution. Our production is crucial and watched carefully by the Executive Chef and all the Sous Chefs. In busy seasons we may have as many as 30 people prepping every day in each of our two commissaries. Tens of thousands of hors d'oeuvres are made weekly, hundreds of gallons of stocks and sauces are prepared, 350+ filets of beef are cut, and hundreds of cases of vegetables are carved, chopped, and blanched. The accuracy, the speed, and the organization of the production are critical to meeting the strict timelines necessary to complete the events. Each area of production has its own banquet Chef with a team of two or three Cooks. Production usually runs from 7:30 A.M. to 5 P.M.; however, during crunch times everyone stays until all the work is done.

## OFF-PREMISES EXECUTION

When an off-premises Chef is scheduled to do an event, he or she usually shows up at the shop three hours before the truck departure. Along with a commissary Sous Chef, the off-premises Chef checks off the event order form line by line to make absolutely sure every item specified is ready. The Chefs ensure the quality of the production items in every respect, from transport containers to quantity. After checking and rechecking supplies, the off-premises Chef is briefed by the Executive Chef about last-minute changes in details, cooking instructions, plate presentation, and so on. After that, the truck is loaded and we're off to the site, where the fun is about to start.

At the scene, the seasoned caterer experiences the controlled chaos as the juice that jazzes the senses. Trucks unload rental tables, gold-leaf china, crystal glassware, and elegant sterling silver utensils. The florist hovers over the head table to demonstrate the elaborate table design, as the captain for the evening places an exquisite table setting as an example for the floor staff. The place is buzzing with streams of people moving in hundreds of directions, doing hundreds of jobs. Can we possibly be ready on time? Once again, between experience, patience, and the gods of catering, the event falls into place.

Off-premises kitchens are usually makeshift facilities that bring a pack list of comforts from back at the ranch. You become a master of Sterno cooking, using hot boxes as your oven to perfect a sit-down meal for 1,000 guests. The off-premises Chef—referred to as the back-of-the-house (BOH) manager—works in tandem with the front-of-the-house (FOH) managers to orchestrate timing, handle logistical issues, and avoid typical obstacles.

The Chef is responsible for leading the staff through the execution of the menu through the entire affair. He or she must coordinate the meal timing while supervising the entire back of the house and communicating with the front of the house. The captain relays the information to the Chef in charge for firing (getting ready) the various courses, yet the Chef must be extremely detail-oriented—for example, he or she must know where the kosher meals are and how many fish or vegetarian entrées are needed—while remaining composed at all times.

The Chef and crew usually work the event for about 9 hours, making for a 13-hour day when all is said and done. A separate crew unloads and cleans up at the commissary so the Chef can head for home. The client usually makes a point of finding the Chefs to thank them for a superb performance; this of course is relayed immediately to the support staff. Another rewarding day is done—but tomorrow is just around the corner.

### MY JOURNEY TODAY

My journey from school goes back some 20 years. When I first started with this company, I never dreamed that someday I would be a main player, a partner in a long-established family-owned business. Once I started working here and was exposed to the company's culture, values, and mission, I knew it was something I wanted to be part of for a very long time. I didn't feel it necessary to move around to different jobs for the sake of moving. I found myself in a position that was both challenging and enjoyable. My responsibilities grew, as did my financial reward. I was clear from the beginning that I required continual growth to feel motivated and creative. Today, although my first love and focus is the food and beverage side of the business, I have responsibilities in marketing, public relations, strategic planning, human resources, and new business development. I love what I do and feel fortunate for the opportunities that have been put in my path. With hard work, dedication, and stability, anything is possible.

## The Reality

They don't call it schlep for nothing! The reality is that it takes a great deal of planning, precision, checking, and rechecking to get a banquet from one place to another without glitches. That said, we are talking about places that provide at least what is known as on-premises and off-premises catering. Most caterers really only do off-premises, but if they have made the investment to secure a permanent facility, they probably do both. Clever entrepreneurs take the time to arrange interesting venues such as boats, museums, private beaches, islands, and even wilderness locations, while some stick to tamer settings like a roof with a great view or a beautiful park. Whatever the venue, if the food isn't prepared there, that means transporting the entire event—usually to a place with limited or even nonexistent facilities. To a great extent, excellence in catering involves maintaining a list of interesting venues that are at your disposal and sometimes even securing an exclusive deal with them.

Logistics, timing, organization, and mise en place are the words that drive the day. There is always a tight time frame for setting up the event, producing it, and breaking it down. The process requires a thorough understanding of what occurs at an event, which will serve as a guideline for your staff, who need to check on-site deliveries and make sure all is in place prior to the eventbeginning. Then your army descends on the function site like bees to honey, each soldier with specific directions and responsibilities. Some of your forces are veterans, others are in training, and many are seasonal regulars. Each must know his or her job and quickly get ready, as you have guests in two hours. It helps if you know the site and all its idiosyncrasies, or at least have conducted a thorough site visit, making notes and anticipating possible pitfalls. Some caterers invest in specially outfitted trucks or elaborate portable work areas and equipment to aid the efficiency of the process, but such an investment is so big that it requires a pretty substantial company to support it. Most caterers still rely on rental equipment that is often not in the greatest condition. Whatever the equipment, you can be sure that it is experience and skill that make the magic of turning an ordinary place into a beautiful event.

## THE GRIND

The truth of the matter is that most catering companies do the majority of their business inside of two four-month periods: March through June, and September through December. This fact alone drives the real grind behind catering. For most caterers, this type of readiness and shutdown relies on a huge network of part-timers who can be available on short notice, or even a rent-a-waiter staffing service, although full-time jobs do exist. You, as a caterer, will probably be able to secure enough intermittent business to support yourself and a small permanent staff. Sometimes it is prudent to maintain a small café, deli, or gourmet store business to help support your core staff. All of these choices can provide a great base income and pay the bills for the much-needed production facility all year round. In all cases you must find a creative way to sustain a nucleus that can be rapidly expanded on short notice, or you will have trouble maintaining consistency of product. This would be the downfall of your business, for one of your most marketable attributes is trustworthiness—the reputation for producing what you say you can. This is especially true because almost all large events—weddings, museum openings, society events, special birthdays—are very important to the hosts. You can be sure that no bride wants a moment of doubt about your ability to pull off her wedding. The idea here is to promise what you can do and do what you promise.

## PHYSICAL ISSUES

The lower back of a caterer, especially one in a small business, is at the greatest physical risk. As we have discussed, catering calls for moving a great deal of stuff, some light, some incredibly heavy, bulky, messy, and hot. Be careful, and teach your staff to lift correctly. The investment in providing them with the proper work tools will pay off in the long run.

## THE BENEFITS FOR YOUR CAREER

As a trusted caterer, you become an important piece of the community. You serve government officials, hospital fund-raisers, local moguls, and many other important people at important events. You must give assurance that family special events are also flawlessly handled and will exceed the clients' best expectations. The fact that you are the food guru will help sustain your business too. Market this position, guard this position, and be the Chef they all believe you are.

Becoming joined to the creative end of local culture confers many benefits and connections that will help you move through life successfully.

## YOUR LIFE OUTSIDE THE JOB

As you probably have guessed, because most of the big business is planned far in advance, a caterer's life is pretty predictable, so you can enjoy the fruits of your labors. This can even mean time to get away, especially when you work a light schedule for part of the year. Much of your job occurs outside the office; you get to move around and are not tied to the stove all day, especially if you are the Chef-owner.

# Professional Organizations

Please see Appendix A for more information on each of the following professional organizations for culinary and foodservice professionals working in catering.

American Culinary Federation

International Association of Culinary Professionals

International Caterers Association

# Interview

**ALISON AWERBUCH, Partner and Chief Culinary Officer, Abigail Kirsch Culinary Production**

**Q / Tell us a little bit about your company and position.**

*A: Abigail Kirsch is an upscale, high-volume on- and off-premises catering company headquartered in New York. We operate five in-house catering venues, as well as a thriving off-premises business throughout the tri-state area. Our annual revenue is approximately $40 million, and we employee over 250 full-time and 500 part-time and on-call associates. My title is partner and chief culinary officer.*

*I have varied responsibilities and have an overview and general accountability for all aspects of our businesses. More specifically, on a day-to-day basis, about 60% of my responsibilities entail overseeing all the culinary and back-of-house operational aspects of the operations. My direct reports include three executive chefs, our corporate pastry chef, our corporate director of purchasing, and our operations manager. The other 40% of my responsibility overlaps with all other aspects of our business, including finance, sales,*

The Feel | A Day in the Life | The Reality | Professional Organizations | Interview | Organization | Education Path Advice

**61**

*marketing, human resources, front-of-house operations, and new business development. Although I am not directly responsible for these areas, I do work closely with our key managers in these areas.*

**Q / Who do you report to?**

*A: I don't report directly to anyone. I have three partners in the business, so I have a lot of accountability to them, especially for the aspects of the business that I'm responsible for. I also have accountability to everybody that works for us, for providing the necessary tools to create the work environment in which they can thrive.*

*And I definitely have accountability to all of our customers and clients, in that we're meeting their expectations and giving them the services and the products that they're expecting.*

**Q / Do you meet with an executive committee on a regular basis?**

*A: Yes. Our executive team consists of myself, my partner, who is the chief executive officer, the two founders of the business, our corporate director of human resources, our corporate director of finance, and our corporate director of operations. We try to meet quarterly. In addition, the same group meets with our general manager from each property on a quarterly basis. Our senior managers have individual departmentalized meetings within their properties.*

**Q / How many venues does your company operate?**

*A: The overview of the business is, we've got five venues that we operate full time and, for the most part, each venue has a General Manager. Our venues include the Light House and Pier Sixty at Chelsea Piers, Tappan Hill in Tarrytown, the New York Botanical Garden, and the Abigail K Yacht. We are also affiliated with numerous historic venues and unique sites in New York, New Jersey, and Connecticut, that can be used for off-premises catering. These include Ellis Island and the Metropolitan Museum of Art.*

**Q / How did you first get into the culinary profession?**

*A: At a young and tender age. I always wanted to go into the culinary field, but I didn't think of it as a profession then. So, after high school, I started out pre-med at the University of Michigan and, after two years of undergraduate school, I didn't like the classes I was taking. I worked in doctors' offices and hospitals to get some experience, and I realized that the medical field would not be anything that I'd be interested in, and that I had no passion for it. Growing up, my family was very into food—all kinds of food, including ethnic and gourmet foods. I always loved to cook as a child, and I followed my mom around. Wherever we traveled, wherever we went, things were geared around food. On holidays, time would be spent in the kitchen preparing food. It was just part of my background, and I loved it. I thought that it was something that I really wanted to do. But I didn't have any practical experience in it, so while I was in college I started working in an entry-level job in a pantry at an upscale restaurant, preparing cold appetizers and salads. I loved it, so I stayed there for two years. Then, after I graduated from* the University of Michigan, I traveled to Europe for several months and worked in catering before starting at the Culinary Institute of America.

**Q/ What degree did you graduate with?**

*A:* I ended up graduating with a bachelor's degree in business. I switched my major after two years and selected classes that might practically help me if I went into the foodservice industry. I thought it would help me in the long run, and if the culinary and foodservice track wasn't right for some reason, I'd have something to fall back on. I always felt that having a broad education in business, no matter what, would be helpful. So after I graduated, I continued to work full time for a couple of years before I applied and went to the Culinary Institute of America. I worked in a hotel and with a small catering company, because in the back of my mind I thought it might be something I'd enjoy. When I started working for the catering company, it was something I really enjoyed and felt this is what I ultimately wanted to go into.

**Q/ As opposed to restaurants?**

*A:* My experience was limited to three or four years then, but catering was something I felt very passionate toward. I liked the creativity, the quality of the food, and the ability to develop new ideas all the time. In the hotel and restaurants, things were much more repetitive. At the time, the food just wasn't as exciting, the service wasn't exciting, the presentations weren't as exciting. And it was just a lot more commercial and a lot more corporate, as opposed to catering. The company I worked for was independent and entrepreneurial, which intrigued me.

**Q/ Where did you do your externship during culinary school?**

*A:* I did my externship at a catering company in New York called Glorious Food, which has a tremendous reputation. It was difficult to get into this externship because they had never taken an extern before, and they had no women working in the kitchen.

It took numerous interviews and hounding the Chef-owner to allow me to do the externship. But I did the externship and loved it. I ended up deferring going back to school so that I could stay at Glorious Food about a year and a half. Every time I thought I was going to be ready to return to school, we were hitting a busy season and the Chef didn't want me to leave. They kept giving me more and more responsibility, so that I was really doing all the tasks and had a lot of the responsibilities that some of the Sous Chefs there had, in running events on my own. I found it challenging and exciting. After a year and a half I went back to college and got my associate's degree, and then decided to look for a job in catering in the New York area.

**Q/ Did you want to go back to Glorious Food?**

*A:* At that point, I wanted to gain as much varied experience as I could. I thought my long-term aspirations were to open my own business. My philosophy, which is a lot of people's philosophy, is to learn from other people, be able to then pick the best from everything you've learned, and make mistakes when

The
Feel   A Day
in the Life   The
Reality   Professional
Organizations   Interview   Organization   Education Path
Advice   **63**

*you're working for other people. And my thought was to work for three to five years after I graduated and then start to gear toward opening my own business, and my thoughts were to work for maybe three different companies within that period of time. So I started to research different catering companies. I interviewed with Abigail, the founder of Abigail Kirsch, and really enjoyed meeting her, in part because it seemed we had a lot of the same characteristics and qualities as far as food philosophy and people philosophy and our backgrounds. For my trial with them, all I had to do was make a crudité basket, and they thought it was the most beautiful one they ever saw. So I think that's why I was hired. I was hired as an entry-level catering chef.*

*Over the first two years, there were three different Executive Chefs who had started, and none worked out. I learned their business quite quickly, although I didn't have nearly enough experience, I think, to be an Executive Chef at the time. Every time one of the Executive Chefs they had hired didn't work out, I was the one that really ran the kitchen, and I would be the one that trained the new Executive Chef. So, after the third Executive Chef didn't work out, I was ready to move on and learn something new. At that point in time, they offered me the opportunity to be Executive Chef, and that was reason to stay because it provided me with a great opportunity and great challenges and a lot more managerial responsibilities. And what transpired after that was that the business continued to grow fairly rapidly and, as the business continued to grow, my job became more and more challenging. So that, every several years when I thought I had somewhat mastered my present responsibilities, something new came on the horizon, and it kept me very challenged and busy.*

*Only five years after I started with this company, the husband and wife who were the founders and one of their sons offered me the opportunity to become a partner in the company. That was something I had to think about long and hard because I knew, if I took that opportunity—and it was a great opportunity—then I would be, at least for a period of time, giving up my aspirations of leaving, my aspirations of having my own independent company, my aspirations of working with other people. But it was just an opportunity too good to pass up, because it gave me opportunities for growth and responsibilities. So I became a partner in the business in 1990, and then that's also when the business was making a real transition and we wanted to start to really grow the business from just off-premises catering to owning and operating on-premises catering facilities. In 1990, we opened our first catering facility and, over the next ten years, we opened four other catering facilities in addition to the off-premises catering.*

*The business had seen tremendous growth in the 1990s, and the last several years we decided we wanted to go in another pattern of growth and are really gearing ourselves up for that now.*

*The business of catering has evolved tremendously over the years. When we started out in the early 1980s, we were one of the first companies that focused on quality food—restaurant-style food and quality service, with service staff trained just as in the finest restaurants. We planned everything from the tenting to the electrical to the special lighting to the floral designs and music. And, when we first started out, that was very unique. Very few offered these services and really were able to execute at the highest level. Nowadays there are a lot of catering companies that offer these services. But we feel what we offer is unique because of our expertise in all aspects of logistics, planning, experience, and meeting and exceeding our guests' expectations. Also, our company's culture is that the customer comes first as well as our associates, and that no matter what we have to do, we're going to do the right thing to make the events great. We have unique, outstanding presentations, and we stay on the cutting edge of what's going on in the culinary world with trends, yet without being too trendy. I think a lot of people don't understand catering, especially on-premises catering, and they think of it as a banquet hall, the rubber chicken. They often don't correlate it to fine dining like restaurant dining. We are constantly changing and improving where there's reason to change and improve, not only with our menus and our presentations but in all aspects of our business.*

**Q / How do you remain on the cutting edge?**

*A: First of all, I don't think you need to remain cutting-edge by job-hopping and working all over. Even though I didn't work with a lot of other Chefs previously, I think I stay cutting-edge by knowing a lot of Chefs in the industry, networking with them, keeping up with professional and other periodicals, dining out a lot, traveling, staying involved in a lot of different organizations that are related to the hospitality industry, and, at times, taking continuing education classes. So that I'm able to observe a lot, see a lot, network a lot, travel, and so on. That's how I stay stimulated and motivated.*

**Q / You hit a certain level in your career where it needs to be self-motivating.**

*A: Self-motivation comes when you love what you are doing and truly enjoy creating and making changes.*

**Q / Can you talk a little bit about your management philosophy?**

*A: First of all, I'm a people person, and I think that's really important because you have to enjoy working with people, teaching people, listening to people, learning from people, having a good time with people. I think that one of the key factors to having a good team around you is being able to really enjoy who you're working with and having them enjoy where they're working, having them feel that they're learning and being challenged. Your team needs to feel that they're being appreciated, motivated, and given opportunities. A lot of times, you can keep them challenged simply by asking their opinion and by communicating to them a lot. So that's a big key to the philosophy.*

*My goal is to be able to promote somebody from within or have them be able to move on and get a job that's of a much higher level because of what they've learned from us. Culture is really important. A good work environment involves treating everybody with respect, having a lot of different programs in place to make people feel good about where they're working, giving them incentives, providing fun things for them to do outside of work. The basic* core philosophy is the same, but obviously with different levels between a line employee and a manager and a senior manager; the way of motivating varies a little bit. But that's the basic philosophy.

**Q/ What are the hours of employment like for people who work in catering?**

*A: Some people think catering is a real part-time type of job. But there are definitely full-time opportunities and, especially if you're in a metropolitan area, there are a lot of large-volume catering companies that offer great jobs with great benefits. Compared to the à la carte industry, what catering allows is definitely a much more flexible schedule. It's not for everybody, but typically each workweek varies quite a bit, because your shift and your hours are dependent on events, so that's different from a restaurant where you have set days and hours. So, some people don't like the uncertainty. But, for a lot of people, the variety of it is appealing.*

*Catering is definitely seasonal, and there's not a region in the United States where there isn't seasonality in catering, and the seasonality varies based on locale. In the New York area, the slowest times in catering are the dead of the winter—January and February—and then in the summer it slows down a little bit—in July and August. Then the rest of the year, business is pretty steady. There's steady work through March to December, and as far as the workweek is concerned, it varies company by company. But for full-time associates, it is typically a five-day workweek. You'll have six-day weeks in the busy times, five days in most of the typical workweeks, and then maybe a three- to five-day workweek in winter and in the summer when it is a little slower.*

*The hours vary based on level of responsibility, but typically we shoot for an eight-hour day. For managers and Sous Chefs, it's more likely to be a 10-hour day. I always tell anybody we're interviewing and hiring for a Chef position that, on an annualized basis, it's about a 50-hour workweek. There may be some weeks they're working 55 or 60 hours in our peak times, and, in our slower times they can work 30 to 35 hours. Basically, 80% of the year our Sous Chefs work a five-day workweek. We really try to keep it to that except for the really extreme times. I've always had the philosophy that if you don't burn out your associates with excessive ongoing work, then they will perform at a higher level, and they will stay working with you longer. This philosophy has proven true in our organization.*

**Q/ What is your biggest challenge?**

*A: The seasonality of it makes the job extremely challenging. The only positive of seasonality is that it gives everybody a little bit of a break. But, from a revenue perspective, from a planning perspective, from a staffing perspective, from a buying perspective, the peaks and valleys make it really difficult to plan and manage a business. Everything you think you've mapped out can change on a dime, so that for running a profitable business, it's extremely challenging to keep quality and consistency. So we always say if we could take our same revenue and have the same volume week in and week out, or comparable volumes, we'd be running a very different business. So that, to me, is always the biggest challenge and the biggest frustration at times.*

*The other difficult thing I'm finding in the last three to five years, and it's more industry-wide, not just catering, is really finding and hiring the right staff. Because whether it's front of the house or back of the house, whether it's management or line, it's just difficult to find professionals who are really committed and interested in working in our industry and learning and growing in the industry. So that all I really want, all I really look for, is somebody who is excited and passionate, dedicated and motivated, and wants to learn.*

**Q/ What do you love most about your job?**

*A: Everything else I love. I love the fact that catering is always changing. No two events are ever alike. The clients vary; their needs vary. The type of event varies. It could be a social event, a corporate event, a benefit event. The menus need to vary all the time, as does the style of presentation. You're dealing with a variety of people, a variety of ideas, a variety of talents, and a variety of ways that they envision their event. It's very different when you have a restaurant and you create your one style or philosophy and people come to you for that.*

*Most caterers have to be like chameleons because, if we only have one style or philosophy, we're going to have a much smaller pie of people that want to use us. But that's what I love about it, because if you're creative, that's what keeps me interested and challenged. It also keeps our staff really interested and challenged; if they had to produce and repeat the same food and the same events day in and day out, week in and week out, then it would be like a factory. So that's one thing I love.*

*I love the fact that we run our business like a tightly run corporate business, but with a lot of entrepreneurial spirit. And I love the fact that I think we're the best of both worlds because, if we have corporate structure, but with the entrepreneurial spirit behind it, we're passionate about committing to change and to perfection and trying to always look at what we can do better. I think that, in a true corporate environment, that's not always the case. And we're able to make the change very quickly. So that, if we meet on a Monday and we realize that there is an issue with an event over the weekend and that we need to change our system of doing something, we're able to get the right team in place to brainstorm it, to think it completely through, and then within a day or two, make a change company-wide. It's a great thing.*

**Q/ What kind of advice would you give a new graduate who was looking to go into catering?**

*A: I think anybody, whether they're interested in catering or another field, should first do some research to find the top five, eight, ten people in that field, because I think it's really important to associate yourself with top people in the field. That doesn't mean it's the people that do the highest volume or the highest revenue, but operations where you think you're going to have the opportunity to learn the most from a quality, operational, and business standpoint.*

*Then, I think it's a great idea to try to set up appointments to see if you can meet with them, just to spend an hour or two to learn and understand what their business is like, and ask the questions that are most critical to you. Especially if you're starting out in your career, you may want to know what the workweek is like, what type of room there is for advancement, what your career path will be like once you gain the experience at this establishment.*

*The next step I think would be, if you're really uncertain whether you want to go into catering or another field, to see if you can trail someone in the kitchen for a week or two, so that you can really get a feel for it. Because I don't think there's anything worse—especially in one of your first jobs—than starting a job and finding out two weeks, four weeks, six weeks later, that that field isn't for you.*

*For some people, especially those without much experience, I feel it's better, in catering, to start to work for a very small company where an owner or head Chef is very hands-on, where you'd be working side-by-side with them, so you'd be able to not only learn the culinary aspects but really start to pick their brain about other parts of the business. In a small company, you're also going to have the opportunity to do a lot of different things. From a culinary perspective, typically there are only a few people in the kitchen, and you have to do everything from making hors d'oeuvres to making stock to helping with dessert.*

*In a larger company like ours, it's more departmentalized, so that you don't get quite as involved in making a variety of items day in and day out. You rotate, so over an extended period you have the opportunity to do that. But the upside of a larger company is that you get great experience in volume production and volume execution of events. You know, once you learn that, and you learn the right ways of doing it, and you learn great organizational methods, then that applies to a variety of industries. So a lot of people who leave catering and enter other segments of the industry feel they bring wonderful organization and discipline to their stations.*

**Q / In catering, there are many opportunities. Can you describe some catering careers?**

*A: Catering is similar, if you look at an organizational chart, to almost any operation in the hospitality industry. Most catering organizations have an owner/ CEO or a General Manager at the top of the organization chart. Then typically the umbrella under the General Manager would be managers in finance, human resources, sales, culinary, purchasing, operations, and maybe marketing. To go the next tier down of management, there are the Sales Staff, Event Managers or Maitre d's, Beverage Managers, Executive Stewards, Sous Chefs, Facilities Managers. The next tier of associates includes Sales, Administrative Support, Captains and Waiters, Cooks, Stewarding and Warewashing Staff, Receiving Clerks, Housemen, and so on.*

*In the back of the house, there is an Executive Chef, Executive Pastry Chef, Executive Sous Chefs, and Catering Chefs. In our company, we have a corporate director of purchasing, and each of the different facilities has a purchasing manager. I'm very strong on promoting from within and promoting Chefs to become purchasing managers. We actually have an opening right now, and we're just discussing it today, and to me what is important in that position is somebody who has a culinary and a food background rather than a traditional purchasing background. It's more and more becoming the norm. Some of the bigger companies now have Master Chefs who, in fact, are in charge of purchasing—for huge corporations. Oftentimes, in the past, a traditional purchaser knew nothing about food and, even if they were good at negotiating bids, really didn't know if the quality was coming in right and if there was more waste in the long run. With culinary experience purchasers can help identify these issues before they ever reach the kitchen.*

*The one position I forgot about is food and beverage management. And that's a position in our industry now that, in a slight way, is being phased out in a lot of operations, because the Executive Chef becomes the food and beverage manager. In catering, there are still companies that have food and beverage managers, depending on the size of the company, so that's another opportunity.*

*One other aspect of catering that's interesting is that there are a lot of opportunities to work part time and freelance to pick up extra money.*

### Q / What trends do you see in the industry?

*A: A trend that is definitely happening is that to be a Chef now, you can't just be a great culinarian. You have to be a very well-rounded person with great business acumen. You have to be a good people person, a good leader, and a good motivator. You have to have human resources skills, as well as be an administrator. You have to wear a lot of different hats. I think the days of somebody being a great Chef but not being able to manage a kitchen, motivate employees, and understand your business are on their way out.*

## ADVICE FROM A RECENT GRADUATE

Name:        Matt Weber
Education:   AAS Culinary Arts, Pennsylvania Culinary Institute, 2002

*Keep your eyes open, mouth shut, and your tail moving!*

# Organization

**Figure 4-1**

Sample On- and Off-Premises
Catering Organizational Chart

# Education Path Advice

**Figure 4-2**

On- and Off-Premises Catering—Education Path Advice

Height of ball indicates importance.

## EXERCISES

1. Go to the website of the International Caterers Association (www.icacater .org). Click on "Newsletter" and read the cover article. Summarize the article in one paragraph.

2. Investigate the names and types of two caterers in your local area. Do they do off-premises or on-premises catering? Are they small businesses, or are they large-volume caterers? Where do they prepare their food? What types of clientele do they serve?

3. Go to the website of *Special Events* magazine (www.specialevents.com). Pick a subject you're interested in and read two articles about it, using the current and past issues. Summarize each article in one paragraph.

4. Compare and contrast the job duties of an executive chef in a restaurant with those of an executive chef in a catering business.

5. Why do you need to have a lot of experience to do a job such as Alison Awerbuch describes in the interview?

6. Would you like to work as a chef in a catering business? Which activities would you enjoy? What parts of the job might be difficult?

7. To see what a large-volume catering business can do, go to the website of Abigail Kirsch Caterers (www.abigailkirsch.com).

8. Using the Career Path Guide (page 2) and Education Path Guide (page 70), map out a possible career and education path for an executive chef in a catering business, including:
   - Number of months in each job
   - Each place of employment
   - How and when you would complete your formal education
   - How and when you would complete skills and competencies

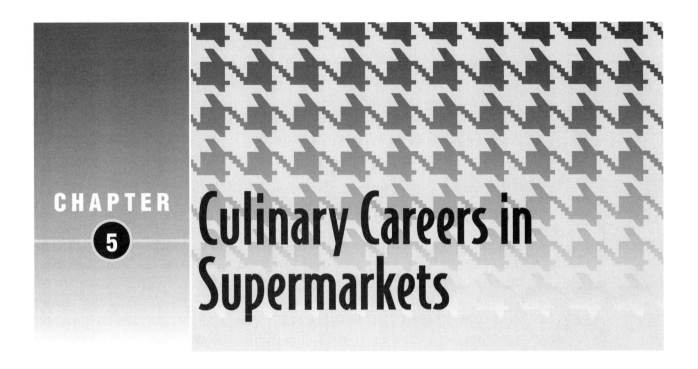

# CHAPTER 5

# Culinary Careers in Supermarkets

## Introduction

**SUPERMARKETS ARE GRAND PLACES** that serve the shopping needs of modern America in one huge area. They stock everything from pastry and breads to authentic Chinese ingredients, and house themed store cafés. From the delicacies of the culinary department, turn left and enjoy the convenience of a day-care center, pharmacy, optometrist, dry cleaning, even a video rental outlet. The conveniences differ from store to store, but the goal is the same: to make a weekly event an experience rather than a chore. With supermarkets offering more and more prepared foods, it's only natural that Chefs gravitate here and take the worry out of "What's for dinner?" and "How do I prepare it?"

## The Feel

### THE BUTCHER, THE BAKER — IT'S 1968

Ring, ring, ring, ring—the sound of groceries being rung up on our family's weekly shopping visit. The colorful foods were wrapped with brand names that stayed printed in my head for many years to come. We were special, growing up in a second-generation Italian family that considered TV dinners a once-a-year treat. Grocery stores were a necessity, with specialty stores still the norm: The butcher, the baker, and the fishmonger all figured in the weekly ceremony of filling the house with food. Mealtime was a ritual with all members present, creating an atmosphere of family bonding, establishing strong and lasting relationships. In the 1960s, 1970s, and even the 1980s, chain grocery markets or superstores didn't provide us with the ethnic or sophisticated foods needed to satisfy our palates. Deli departments were mediocre at best, with industrial salads and unappealing displays of convenience food. Breads were marshmallow in texture, bland in taste, with little or no selection. The cakes were simple; inexpensive ingredients such as shortening replaced butter, and canned fillings were added as a cheap afterthought in a crisis situation.

## THE SUPERSTORE MAKER

In a blink, the focus moves forward to 1997. "Am I crazy?" I ask myself. "Am I making the biggest mistake of my career?" I am moving my family to a new home so I can work at a supermarket. We all know the trend is changing and cooking is becoming America's number-one obsession, but are consumers really ready to be educated? Do they want to know the difference between a snap and a snow pea, between summer and winter truffles? Do they want to eat exotic baby greens? Do they care how to make a béchamel sauce for macaroni and cheese? My heart throbs with apprehensive excitement. The two gentlemen who recruited me had convinced a staunch hotel Chef that the American food market was the next explosion in our industry. But was I ready for a uniform of half suit and tie, half Chef whites?

I would be the liaison between field and office, preparing culinary programs, developing recipes, doing customer relations, holding photo shoots, designing a magazine, and implementing the vision for growth shared by the top executives in the company. I was up for the challenge and believed this career change was the right move. It took about a year to get my feet planted and our philosophy established. We were all on the same page, from the owner to the store attendant, striving to be cutting-edge superheroes. In one year, we became the envy of the industry, and my title was Operations Director of Food Preparation. We supplied a package to our clientele with all the accoutrements, from cut mise en place, marinated and stuffed proteins, cooking instructions, secret sauces, the right cookware, wine to serve, desserts best suited, breads, and even after-dinner chocolates. We believed the more we educated the consumer, the better we would look—a philosophy that has been successful.

## A Day in the Life

Picture a lavish, well-lit store with aisle after aisle of farmers' market produce, extraordinary fresh fish and meat counters, a spectacular selection of domestic and imported deli meats and cheeses, four varieties of our own brand of extra-virgin olive

The Feel | A Day in the Life | The Reality | Professional Organizations | Interview | Organization | Education Path Advice

**75**

oil culled from different Italian regions, and spoils such as Russian beluga and fresh black truffles. These treasures lie alongside the mundane necessities of everyday life such as milk, bread, paper towels, and toothpaste. Now take a tour of our abundant salad bar, Chinese buffet, bakery/pastry shop, and sushi station; admire our restaurant-quality open kitchens. Seven years ago, prepared foods were just home replacement meals. Put back wrong terminology—unglamorous prepared complete dinners craved by the working yuppies of the mid- to late 1990s. Part of the reason for the new mega-markets was the rebirth of the professional culinary regimen I was hired to implement and orchestrate to a level of excellence. The attraction for me in coming to work for this supermarket company was its unmistakable commitment to hiring restaurant-quality Chefs to create restaurant-quality food.

## FIRST ON TODAY'S AGENDA

Conference call with all regional Chefs. It is important when you multitask to be in contact with your core staff who are in different locations. The only way to success is constantly educating your people and encouraging a trickle-down effect of information and attitude, mixed with periodic touring of each location. You hire professionals who share your passion for quality and perfection. The conference call lasts about 45 minutes, recapping newly launched programs and working out potential obstacles to success.

Today I am in the Chef whites part of my uniform. For the next three days we will be shooting (photographing) finished dishes for our monthly magazine; this gives me the opportunity to get my hands dirty, which I love and miss the most. Preaching philosophy and instilling passion in my staff is a full-time job, both rewarding and satisfying. "Lead by example" are the words that have echoed through my head for most of my career. If I want my staff members to wear a hat, then I must wear a hat. If I expect an organized workplace, then I must be organized myself. The Chef is at the helm of this vessel; he or she is the mentor to all staff members, and they watch every move. Many of them strive to become Chefs themselves, which is a compliment to one's leadership as well as a responsibility.

## THE MARKET CAFÉ

The culinary brigades in food markets have been instrumental in reinventing retail food cases. Concepts reflect those of the *Klienmarket* in Germany; the downtown street markets in Tokyo, Japan; the Peck Market in Milan, Italy; the *Mercato* in Valencia, Spain; and the quaintness of an Italian trattoria. We as supermarket Chefs take pride in maintaining the integrity and the authenticity of the foods we prepare. Other stores offer cheap imitations; however, we are committed to educating our customers. Our reputation is built on trust and loyalty, on delivering a quality product few can match.

Food prep is the buzzword, the core of the Executive Chef's responsibilities. Anything related to a food product that requires preparation is supervised by our culinary department. The Perishables Manager, who mainly handles meat, fish, poultry, and produce, relies on the Executive Chef's expertise to provide prepared oven- and grill-ready foods that enhance sales from their elaborate display cases. Such delicacies include fillet of sole stuffed with Jonah crab; macadamia and horseradish–crusted salmon; spinach, dried tomato, and fontina steak roulades; mustard and rosemary–marinated pork tenderloin; and a delicate lemon sole rolled around a scallop, shrimp, and dill mousse.

Every store has a repertoire of established recipes that are logical to implement, possess a refined appearance, offer superior taste, are affordable, and supply balanced nutrition. The hot and cold stations of the café are an integral part of the daily operation. Thick-piled sandwiches are made on a variety of breads with a vast selection of creative toppings, cheeses, and spreads. Sushi hand-rolled to order may be your desire of the day. Select from tuna, fluke, smoked eel, giant clam, pulpa (octopus), sea urchin, or the American tradition, California roll. Next in line is the authentic Chinese buffet serving our famous Chinese sesame chicken, pork with scallions and oranges, sirloin strips with Chinese long beans and oyster sauce, quick-fried vegetables with edamame (fresh soybeans), tofu and jasmine rice, or the picture-perfect jumbo shrimp with snow peas, peppers, and napa cabbage—all packed in the customary Chinese quart and pint containers. The local Tong Hoy restaurant had better get ready to rumble.

## PASTRY/BAKERY

The bakery/pastry shop completes the shopper's experience. World-class hand-crafted pastries and cakes are designed with masterly skill and the best of ingredients. The selection of baked goods has something for everybody. Notice the freshly baked artisan breads, authentic buttery breakfast Danish, scrumptious oversized muffins, and assortment of designer cookies that look as if they're straight out of European bakery cases. Luscious caramelized banana and white chocolate brioche pudding, sinful Dobos torte, gorgeous raspberry chocolate mousse cake, mouthwatering chocolate caramel hazelnut tart, delectable chocolate coffee cupcakes, and a brilliant linzer torte in three flavors are just a few of the sweet endings you can get here for the finale of your fabulous meal. And then the miniatures! Hand-filled candies, ten kinds of fudge, tiny pastries, and many more classics are executed with the same skill and devotion as all the food preparation. I was taught a long time ago to savor a little of a fabulous treasure, then to consume a lot of something mediocre, which sums up how to handle the temptations of this department.

## RETAIL LINE

Next stop is the retail line, a bountiful presentation of quality selections to replace a home-cooked meal; note particularly the comfort-food standards like meatloaf, mushroom gravy, pot roast with root vegetables, and fried chicken with Yukon gold whipped potatoes. The next level supplies a more elegant evening and features the Chef's more adventurous choices. Move down the line to the pumpkin ravioli with citrus-sage butter, wild mushroom and three-cheese risotto, grilled lamb tenderloin à la Grecque (olives, tomato, cucumber, and mint), or grilled pork loin with roasted baby vegetables and apple chutney. Then hit the spread of accompaniments—salads and side dishes that are as delicious as they are enticing. Fabulous platters of these appetizing selections give shoppers an abundance of choices such as broccoli rabe with roasted elephant garlic, eggplant rollatini with fresh mozzarella and ricotta salata, or the pear provolone salad with walnut vinaigrette.

## THE CLOSE OF ANOTHER DAY

The creative opportunities for a supermarket Chef are endless, as is the tremendous customer respect for your many years of cooking experience. It is a great feeling to know you have a new class of students (the customers) interested in your advice, knowledge, and abilities. As a supermarket Chef, you can endlessly improve your skills by self-motivating continuing-education classes; you can spend a few weeks scattered throughout the year to "stadge" (apprentice) at respected restaurants that share your passion for exchanging ideas and information. These are some of the great advantages in being a Chef in our changing world of food.

Today passed like a twister in Kansas, fast and furious. The energy is so positive and the team is so motivated that I feel honored to be around this dedicated group. It is a wonderful feeling to know I am at the peak of my productivity, the prime of my career, and to see I am able to make a difference.

# The Reality

Working as a Chef in a supermarket is not for the inexperienced. On the contrary, it is for those who have many varied experiences and a thorough understanding of the culinary profession. Supermarket Chefs need an unlimited repertoire of dishes and cuisines to support their production. They must focus on sanitation training and Hazard Analysis and Critical Control Points (HACCP) standards, to prevent mishaps; they must distinguish between hot foods to be held hot and home meal replacements (HMR), which are designed to be reheated. In short, this job requires a top-notch pro with quality skills and training.

Note that this is a corporate atmosphere, and chances are that the Executive Chef will be called on to participate in much more than just food production. The Chef must be savvy in the areas of public relations, marketing, business planning, and goals, and must be able to present his or her ideas. Just look at the organizational chart in Figure 5-1, and you will get an idea of all the people the supermarket Chef interfaces with. As the Executive Chef, you will be involved with many types of professionals and expected to have the skills required to interact with sharp corporate businesspeople. You must have strong management skills as well, because your staff will be large and made up of various types of people, including salaried and hourly employees. That is not a negative; it merely indicates that the motivators for your staff will be varied and that you should expect to use many methods to achieve the goals you have set for your business.

## THE GRIND

To a great extent, this job has so many facets that there literally is no grind—it is more like a run! You face different issues from day to day and have a wide variety of challenges to conquer. The opportunities are endless and the creativity motivating as you devise new menus, items for home meal replacements (HMR), and store demos. You may even do some teaching in your on-site show kitchen. Your life is busy and interesting at every turn. Of course, you must regularly deal with the financial aspect of your enterprise and must monitor the processes you have implemented to make sure all is well and profit is generated. You must keep communication flowing to your department heads and motivate them to succeed.

## PHYSICAL ISSUES

Mental stamina is the primary issue in the supermarket; the real physical pain was felt prior to this job, while you built your cooking experiences. Of course, this is an on-your-feet job, so your feet and legs will speak up once in a while, but that comes with the territory.

## THE BENEFITS FOR YOUR CAREER

The broader market offers discount groceries galore, and you know how much that means if you have a family. But a career like supermarket Chef, which offers so many challenges and opportunities, is always giving you a new light to follow. You can take advantage of extensive networking and get to know a completely different sector of vendors, manufacturers, and grocery specialists. There is enormous support for you to take continuing education or professional development courses. Mega-markets are committed and dedicated to developing well-educated, inspired professionals. The trek is enlightening and built on the sound culinary base you developed in the early years.

## YOUR LIFE OUTSIDE THE JOB

You should be able to succeed as a supermarket Chef and at the same time enjoy a high quality of life with your family. Of course, work demands rise at busy holiday times, the occasional ramp-up for a new store, and the promotional rollout of new products. You must travel to headquarters for meetings and planning sessions, and make the occasional trip to check out competition.

## Professional Organizations

Please see Appendix A for more information on each of the following professional organizations for culinary and foodservice professionals working in supermarkets.

American Culinary Federation

National Restaurant Association

Research Chefs Association

## Interview

**JIM SCHAEFFER, Operations Director of Food Preparation, Wegmans Food Markets, Inc.**

**Q / As Director of Food Preparation for Wegmans, to whom do you report, and how many people report to you?**

*A: I report to Jim Berndt, the Senior Vice President of prepared foods for Wegmans. I have three direct reports in the office. One person does all the nutritionals, labeling, and costing for all of our products. I also have an Executive Chef of Asian cuisine. And another training Executive Chef also works for me. We work directly with all of our Executive Chefs in the stores, doing the openings, remodels, and training, but we also set the standards and specifications for our programs, such as packaged sandwiches, packaged salads, seasonal soup cycles, and different food festivals like Mardi Gras and Oktoberfest. It is very exciting.*

*We do a lot of different things. For example, we have done a dinner at the James Beard House. I spearheaded that event along with our Chefs from some of our different stores. Also, last week we did an event in our restaurant, Tastings, with David Bouley for two nights. Tastings is our only restaurant, and it is attached to our Pittsford store in Rochester. It has an open kitchen and serves lunch and dinner.*

**Q / What is *Wegmans Menu Magazine*?**

*A: It's a publication that is put out four times a year. It's designed for our customers and gives them menu ideas, recipes, and cooking techniques such as pan-searing, braising, or roasting vegetables. We're in our fifth year of publication. It's been a great success.*

**Q / Can you give me a little bit of a typical day, or if that's not enough, a typical week of what you would accomplish?**

*A: The beauty about it is that it is very, very different all the time. Today I did everything from working on a design for an outside cooking kiosk that had a fryer, a grill, a beverage component, and a cash register, to meeting with a soup vendor, to meeting with people with Consumer Affairs about proper seafood temperatures that will be communicated to customers. Tomorrow I have employee evaluations, a financial accounting meeting, which is a couple of hours, and a brainstorming meeting about an upcoming store managers' event where we cater for 120 people. The final meeting of the day is about a central production facility we are planning to open in 2006. We currently have a cook/chill facility where we do our own sauces, soups, and purees. There is a fine balance between what we do outsource centrally and what we do in-store. I also of course spend time in stores, cooking, fine-tuning programs, and speaking with our employees and customers.*

**Q / Do you go into core recipes?**

**A:** *Exactly, and I'll do all the photo shoots for that and then we will do product rollouts, division by division, to make sure everyone gets it, tastes it, feels it, and then they can go back and be the cheerleader for their store.*

**Q / How many stores are there?**

**A:** *There are 67 stores, and we grow by two to three stores a year. This year we have two stores that we are going to open, and next year we have three on the boards. The stores are split into regions, and we have Senior Vice Presidents who are responsible for anywhere from 5 to 12 stores.*

**Q / Where did you go to school, and how did your career unfold?**

**A:** *I got into this business by accident. I went to a restaurant, the best restaurant in my hometown in Wisconsin, to work as a busboy with a friend of mine. From day one they put me into the kitchen, and I've been working in the kitchen now for over 30 years. All my friends were playing football and I was cooking, and that was not an attractive profession 30 years ago. But I needed to have a job, so I did it, and then once I got out of high school, I wasn't old enough to get a job in a factory so I was kind of forced to stay in foodservice, and luckily for me I began to enjoy it.*

*Growing up, I was a poor high-school student because I didn't apply myself. I always thought that going to college was definitely out of reach; my family couldn't afford it. It was not until I worked a couple of full-time jobs and I started accumulating money that I figured, the only way I am really going to excel at this is if I get a formal education. I used to think it's all going to be experience, but it's really those classroom settings that are equally important. I wouldn't be where I am if I hadn't had that. I'm sure I would have done well, but not as well, without the formal education.*

*I had cooked for five years before I went to culinary school, which I think for me was a big advantage because I was a little more mature, and I was paying for the school by myself. So when I got to school I was really focused. I spent a lot of time in the library and the audiovisual area because it was my money. I really didn't have time to mess around. It wasn't my first time away from home, because I had lived in Chicago for three years by myself.*

*Sometimes students go there right out of high school, and their focus is not necessarily there yet.*

*I went through my two years at culinary school, and I did my externship at the Fairmont Hotel in Philadelphia, where I worked with many talented people. Although I was only there for a short period of time, I really grasped a lot of knowledge and made a lot of good contacts. After graduation I went to the Playboy Hotel and Casino in Atlantic City, and I opened the hotel as the Banquet Chef in March 1981. The hotel had 18 Sous Chefs, 220 cooks, 190 stewards. Back then, the hotel had $36 million in revenue, which was significant. There were nine restaurants plus banquets. Because of the volume, I got a lot of exposure and a lot of experience that paid off later on, in my years with Hyatt. In 1983 I became Executive Sous Chef at the Playboy.*

*In 1986 I decided to learn more in another job. I was fortunate that Hyatt Hotels gave me the opportunity to become an Executive Chef at 27 years old. I went from a hotel where expense was not a problem to a 400-room union property that had difficulty paying the debt service every month. From a staff recruitment standpoint, I was competing against hotels and restaurants that could pay their employees more money than I could, so I had to not only provide a paycheck but also provide mentoring, upward mobility, and a great work environment. I was really fortunate to develop a great team of people and promote a lot of people.*

*In 1989, I was promoted to regional Executive Chef at the Hyatt Regency in New Orleans, a 1,200-room convention hotel. I was in that position for eight years. Then in 1997, I moved on to Wegmans Food Markets, Inc.*

**Q / What is your management style?**

*A: That has always been one of my strengths. I lead by example. I work hard because I want the people to work hard. I wear a hat because I want them to wear a hat. When the stuff is hitting the fan in the restaurant, in the banquet, wherever, my staff knows that I'll be there. They can always count on me and, therefore, I can always count on them. That's huge. No matter what you are doing, you lead by example. The bottom line is that it is going to be up to them whether or not they do their best.*

**Q / What do you find was the most valuable part about your education?**

*A: No question about it, it was skill development. If you don't have a really rock-solid foundation, I think you'll have problems. There's a lot to learn in this field, which makes it a very humbling career.*

**Q / Why do you feel that you excelled in your career? What are some of your best attributes?**

*A: I've always prided myself on being a complete package. There's always going to be people who can cook better than you, but I think the ability to manage a team and mentor people is very important. I'm a very consistent person; I'm methodical. You have to be a professional every day. I've also always been pretty humble and modest. I'm usually one of the first people here and one of the last to leave. There are days that are very, very demanding, and I meet those challenges.*

**Q/ What are the things you like the most about where you are now?**

*A: I like it because we pioneer a lot of things. I like it because of who we are. We are always on the cutting edge on whatever it is. We are the envy of the industry. That's pretty exciting to be part of that team. Also, I've been blessed with the ability to explore a lot of things that I only read about in the past. I've traveled to Europe, Japan, Thailand, and Hawaii to check out food and food trends.*

*It's a great company. It's the ninth-best company in America to work for, and we're striving to become number one. Chances are pretty good we will accomplish that. Everyone just operates at a very high level of integrity. We want to deliver the best product to the customer, the top of the line, a great product at a great value.*

**Q/ What advice would you want to give young aspiring Chefs?**

*A: Stay hungry, work hard, stay focused, be a good person and a role model. Lead by example. Eventually cream rises to the top. It's always going to be that way, always.*

## ADVICE FROM A RECENT GRADUATE

Name:         Adam Volk
Education:   AOS Culinary Arts, Culinary Institute of America, 2006

> *Take [school] very seriously. Treat it like your job and take pride in your work.*

# Organization

**Figure 5-1**

Sample Supermarket
Organizational Chart

## Education Path Advice

**Figure 5-2**

Supermarket—Education
Path Advice

Height of ball indicates importance.

# EXERCISES

1. Visit the Wegmans website (www.wegmans.com). List six features on the site that could be useful to home cooks.

2. Compare and contrast the job duties of an Executive Chef in a restaurant with those of an Executive Chef in a supermarket.

3. Why do you need to have a lot of experience to do a job such as Jim Schaeffer describes in the interview?

4. Would you like to work as a Chef in a supermarket? Which activities would you enjoy? What parts of the job might be difficult?

5. Visit several local supermarkets and compare the types of prepared foods they offer. What food-borne illness concerns are possible with supermarket foods?

6. Visit the website of Whole Foods Market (www.wholefoodsmarket.com). What distinguishes this chain of supermarkets from most others? On their home page, click on "Products," then on "Catering." Print out a catering menu for the Whole Foods Market closest to where you live. What clientele are they trying to attract with their catering menu?

7. Using the Career Path Guide (page 2) and Education Path Guide (page 84), map out a possible career and education path for an Executive Chef in a supermarket, including:

   - Number of months in each job
   - Each place of employment
   - How and when you would complete your formal education
   - How and when you would complete skills and competencies

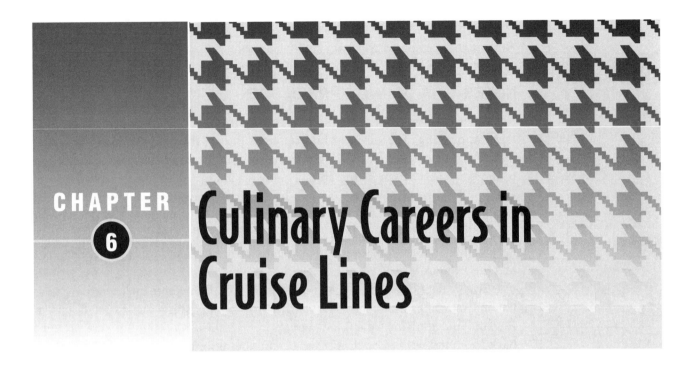

# CHAPTER 6

# Culinary Careers in Cruise Lines

## Introduction

**CRUISING IS ONE OF THE FASTEST-GROWING SEGMENTS OF THE TRAVEL INDUSTRY.** Because cruising appeals to a wide range of tastes, lifestyles, pocketbooks, and interests, cruise ships have become a popular vacation destination. The typical cruise ship carries 2,000 passengers and 900 crew members. Popular places to cruise include the Caribbean islands and the Mediterranean Sea. Many ships today offer vacationers luxurious, full-service spas and health club facilities; specialty restaurants and dining options to satisfy any taste; professionally staffed children's facilities; sports activities from golf to rock climbing; Internet centers; and many other services designed to cater to vacationers of all ages. Working as a Chef on a cruise ship is an unforgettable experience. Read further to learn about culinary careers in this distinctive work setting.

## The Feel

**SURMAGE** — "Listen up!"—Always said before calling the order in to the Station Chefs)

**ARBOYER** — Usually the Sous Chef or Chef tournant who calls the orders to the Chefs of the different stations

87

GLORYHOLE — The Steward who looks after the crew quarters

BULKHEAD — A separating wall, such as a cabin wall, within a ship

## NOTES OF A YOUNG APPRENTICE

The first day of work was completely overwhelming, while at the same time electric and exciting. The throb of huge engines pulsed through the massive vessel, and I heard the unusual clicking and clacking of pipes as the bulkheads began to move to the rhythm of the ship. The tugs took up the strain on the mooring lines, pulling the vessel into the main water of the dock area where we had been berthed, lining us up to go down the channel and into the open sea. The mind-awakening sound of the ship's horn penetrated deep into my skin, a wonderfully deep bellow that acknowledged the small helpful tugs. I felt a surge of excitement and the exhilaration of adventure.

It was a new beginning, being away from my homeland, meeting new people, and starting a career I had yearned for since visiting the finest hotels in England and applying for a three-year apprenticeship. My thoughts were brought back to reality as the strict arboyer began yelling out in deep, confident tones, "Surmage!" We stood at attention at our stations, ready to accommodate his every wish, hoping our presentations would pass his inspection as well as that of the Maître Chef de Cuisine. The variety of mise en place we needed to prepare was quite overwhelming to a newly hired beginner. The secrets: Don't speak unless spoken to, don't stop working unless relieved, and don't make any mistakes.

The night before we sailed had been a weary one. All the apprentices climbed into their bunks in the cabins assigned by the Gloryhole Steward. Each cabin held four men, two bunkbeds, two closets, and two drawers in which to store four months' worth of goods. That was going to take a bit of organizing for four young inexperienced mates who worked day and night and who would rather play cards in their spare time than clean or do laundry. At 2 A.M. we were up on the deck, cold and tired, for a mandatory safety drill emphasizing evacuation in case of an emergency. Sleep-deprived, we slumbered until the crazy Gloryhole Steward started his banging, hooting, and hollering rampage at some ungodly hour, rallying the troops. The moans of my coworkers over how early it was and how tired they were could barely be heard over the banging of our illustrious Steward. But off we went to the head (bathroom) to get ourselves together, knowing full well the Chef was in the kitchen, ready to ride us for yet another day.

First and most important, we were inspected by the Maître Chef de Cuisine, then the Executive Sous Chef, then the Chef de Parties, and so on down to the Sous Chefs and the third Commis for cleaned, pressed uniforms, clean fingernails, a smooth shave, and mirror-shiny shoes. I was just a kid, 18 years old—what did I know or care about being inspected in the morning? After all, I didn't join the military. I came here for a career in cooking. Not until years later did I appreciate the military-style apprenticeship I received, which helped mold me into a professional.

I remember the first day I climbed the steep gangway aboard the liner. She was a beauty, sparkling with nearly three acres of exterior glass and containing a collection of five-star hotel activities and entertainment. The grand plaza, referred to as the atrium, was a huge three-story open area that seemed even more magnificent as I studied its details. Now, it's 2 A.M., and I am assigned to the midnight buffet, which allows me to sleep in an extra hour if I can survive the deafening sounds of the wake-up call. I've decided to start a journal not only for my culinary notes and recipes but also to describe the exquisite ambiance, the people, and the profession to which I will be exposed on this grand sailing ship. I must take a minute and write home, for they will never believe in the small town I come from the wonders I am exposed to daily. I am privileged to be aboard this mighty vessel. The experience will surely be well worth the journey.

## THE ACTIVITIES AND AMBIANCE

The impression of luxury begins when guests board the ships. Curved hallways invite exploration and curiosity; art and antiques make guests feel comfortable, yet they surprise and delight. Fine art is an integral part to the cruise experience, giving a feel of the classics with a touch of contemporary style. Guests are greeted as they arrive by an eager, well-trained professional staff passing savory hors d'oeuvre and vintage champagne to mark the start of an unforgettable experience. Imagine a full-service day spa that offers acupuncture for pain and stress management. Picture playful options such as a rock-climbing wall, jogging track, basketball and volleyball courts, ping-pong tables, and a nine-hole miniature golf course. Night owls can cruise to the martini lounge of the full-blown casino. The kids are all tied up with their own activities—scavenger hunts, sports tournaments, and arts and science workshops. The teen hangout is a nightclub complete with soda bar, dance floor, DJ music, and karaoke rooms—or teens can visit a video arcade packed with every game they ever wanted to play. Professional productions for the whole family to enjoy are staged in three theaters—and don't forget the comedy club, showing live performances from well-known television comics.

The ship has three pools: an indoor pool with a skylight roof, perfect for a cool evening dip; a spa pool located next to the gym, which also houses a beauty shop, massage center, sun tanning, and other health treatments; and the main pool on the sun deck. This one is adjacent to two large dining rooms serving a marvelous, delectable, and well-monitored midnight buffet. Tantalizing food stations satisfy the whim of every guest, whether off to the late-night party places or turning in before a busy day. Besides the elegant à la carte dinner that is available every night, authentic Thai, Mediterranean, sushi, and tapas restaurants offer the same quality and attention to detail the other food outlets are expected to maintain.

I had always heard of the phenomenal Garde Manger work for which cruise lines are so well known, but to witness its grandeur in person is amazing. Grand buffets are the highlight of the cruise ship experience, not only to consume but to observe and experience. To have creative control of these extravagant food stations is a privilege. Although the art of Garde Manger has lost the intricate style of the classic kitchens of great hotels and country clubs, the fresh approach taken on cruise lines has revitalized these classic skills. Customers now expect a certain Garde Manger skill level on cruise ships.

## THE KITCHENS

Thirty-nine hundred assorted muffins, 400 pounds of exotic fruits, 1,600 pounds of beef, 1,430 pounds of game and poultry, 1,170 pounds of potatoes, 960 pounds of bananas, 910 pounds of ice cream, 852 pounds of shellfish, 551 pounds of butter, 390 pounds of bacon, 550 pounds of vegetables, 306 pounds of veal, 650 pounds of fruits, and 104 pounds of salmon are just some of the provisions used daily on this cruise ship. The space of the main kitchen is tight yet accommodating, as long as everyone does his part to stay clean, neat, and organized. Everything is slick stainless steel. All the cabinets have covers so nothing slides out when the weight of the vessel shifts. The cooking battery resembles that of any hotel-sized kitchen and includes everything from a Swiss brazier to double-stacked convection ovens. This is a purely from-scratch kitchen, except for some of the meals executed for staff members. Deliveries are made by helicopter, so you must order well in advance, which leaves you little room for mistakes! You must use whatever food products arrive, because returns aren't an option.

In addition to the main kitchen, satellite kitchens serve the other restaurant options available to guests. You as the Chef must be in constant touch with these satellites to ensure that their quality meets the same standards as the main preparation area.

The full pastry kitchen presents the same space challenges as the rest of the food prep areas. Stunning afternoon teas feature elegant pastries, intricate tea sandwiches, and assorted scones as an impressive culinary interlude before dinner service. The cool afternoon breeze and the spectacular horizon are the backdrop for these popular afternoon events. Keep in mind that guests are on a revitalizing vacation with a multitude of activities, so the rich pastries usually passed up at home are a well-deserved treat on board the ship, and they must exemplify the standard of excellence the cruise demands.

## A Day in the Life

Dining aboard these magnificent vessels hasn't changed since their inception. It is an event that lasts from the time you wake until you collapse at night, not to mention the 24-hour first-deck pizzeria. Imagine unwinding at an after-hours place directly across from the casino with selections such as gourmet goat cheese, artichoke and veal sausage pizza with organic baby greens, roasted garlic bread, and signature calzones. There are so many styles and concepts of restaurants that you are sure to find a cruise line that meets your needs. From casual to elegant, the experience is meant to be unforgettable.

## THE ROUNDS

Let's tour these intriguing outlets as they reveal the legendary culinary rituals for which most cruise lines are famous. It's 4:30 A.M., and I feel as though my head just hit the pillow; I worked a mere 17 hours yesterday. This really isn't too bad, considering I get three months on and a month off to recoup. Besides, I have little or no expense, so I can save, save, save. It is Day 2 of a 14-day cruise to the Panama Canal and the Chilean fjords. I have been fortunate to visit many wonderful places since I joined this cruise line five years ago.

The first place on my rounds is the main kitchen, which is the core of everything that takes place. One constant is that for the most part your crew is reliable, and unless someone falls ill, everyone is present for work. Preparation for breakfast is in full swing, with the morning Sous Chef calling out checklist and preparation details to the cooks and Commis. "Morning, Chef," they shout with respect. They know what it takes to attain this position. Working on a ship quickly shows what kind of manager you are in terms of organization, philosophy, cooking skills, and production schedules. You can't fake anything in this environment, and if you aren't right on with your planning, your crew will lose all respect as they run frantically around the kitchen like a disrupted nest of ants.

The breakfast Arboyer barks out orders as the competent waitstaff ensures that all items are being put together in a timely fashion. Five seasoned breakfast cooks move in sync as they flip eggs, layer short stacks, and fold generously stuffed omelets. All quiets down for the most part by 10 A.M., at which time a limited crew remains available for stragglers as the kitchen scrub-down begins. Everything must be cleaned and organized in preparation for the next meal period. You hold a brief recap with the Sous Chefs as the crew continues the endless task of meticulously preparing the next feast. Your creative assistance is needed for several VVIP suite functions involving the captain and some of our most prominent guests. You must spend a patient, well-focused moment reviewing the final product orders, staffing issues, and menu delicacies for the evening.

The ship is at full capacity with six restaurants and smaller grab-and-go outlets on every deck; these are designed to satisfy the guest on the move with an overstuffed sandwich on a ciabatta, handcrafted sushi, or salads with an abundance of toppings. The crew comes back from a short power break to plow through the festivities of the evening. All restaurants are booked with a ship full of influential clients. There's a burst of energy from everyone, placing the finishing touches on the exquisite Garde Manger work or the final chocolate filigree on the Pastry Chef's petit fours. The more experienced Sous Chefs are mentors to the younger, less-experienced cooks and Commis, who shine with pride as they note their accomplishments of the day. You look around and know this is why you are in this business: the rush, the enthusiasm, the passion, the commitment, and the yearning for creativity. As the guests arrive in their beaded gowns, black ties, and tails, you can hear classical music throughout the ship's corridors, creating a feeling of wealth and sophistication. The crew is rehearsed and ready, needing little or no instruction to deliver the superb dining experience one would expect on this vessel.

## THE END OF ANOTHER DAY

At last, I have a minute alone atop the first deck to reflect on the accomplishments of the day. The cool ocean breeze eases the kitchen's heat from my face as I gaze over the same horizon witnessed by great sailors many years before. I remember hearing when I was growing up that people are drawn to water because our bodies are made up of about 75% of this magical liquid. There is a calmness when one is completely surrounded by water. The challenge of this 975-foot, 115,000-ton mass of molded beauty is intense, considering the diversity of clientele, the level of quality, and the many dining options. The days are long, hard, and demanding, especially in the beginning, when the crew is making last-minute adjustments, but the benefits combined with the gratitude are priceless. These floating resorts with their impeccable service create an atmosphere, a memory, an experience guests will remember forever. This is the essential reason I spent my early days training to become an accomplished Chef. The hum of the midnight air whispering against the water is mesmerizing as my thoughts drift into tomorrow's activities. Rest is first, before I dive into another full-scale day of events. One last look around, and I say to myself, "What a great day." Then I think, "No, what a great life."

# The Reality

The cruise ship is a floating amusement park, resort hotel, and dining mecca, so as you may imagine, it is no small task to get it all to function as advertised. In your charge is room service, multiple food outlets, and the special event functions. Don't forget that labor relations, product procurement, and standard operating procedures are included in that mix. You'd better have your act together, or it will not come together; this is not a fake-it-until-you-make-it kind of job. You are master of the galley and must perform flawlessly to keep your young charges in line and respectful of your orders. This is perhaps one of the toughest leadership positions in the culinary industry. It is important to keep in mind that you are going to work side by side with the same people for three months, night and day. You will eat, drink, and sleep in the same tight quarters. You must maintain order as well as morale. It is a tricky spot, but given your years of experience and well-polished abilities, you can do it. Oh, by the way, I hope you don't get seasick!

## THE GRIND

Attendance is recorded, and production has begun. The day's events are demanding your attention, as are the details of supply, maintenance, and special events, which must be discussed with the Chefs in charge. It is the start of another day on board, and we have a port of call at which we restock and reevaluate our needs, then perhaps get a few moments to visit land. There is a grand lobster bake tonight, and then a dance and midnight buffet. You have a great love affair with this operation—the excitement, the globe roving—although it is also a bit lonely, even when you are surrounded by comrades and your trusty staff. The routine can be monotonous, but as you know, this is true to some degree in every foodservice career. However, the monotony can be magnified by the space restrictions, the extended never-ending workdays, and the familiar ports of call.

A great benefit of working the cruise lines is that, while you are on board, you really incur almost no expenses. In addition, some good tax loopholes are available because your earnings are usually based at sea. All in all, the lifestyle is unusual but rewarding, once you acclimate to it.

## PHYSICAL ISSUES

Your feet and legs tire as the sway of the ship tends to strain the muscles even more than the typically long working day. You must get used to the confinement of the ship. Even though it is large, it is limited; there is no town to go out on at night, and there is no country to take a drive in. You must also grow accustomed to the constant motion. This is easy for some, but not so easy for others!

## THE BENEFITS FOR YOUR CAREER

You can gain a plethora of knowledge all in one place on a cruise ship, given your exposure to dedicated professionals, a range of food styles, and elaborate buffets. You can learn from the best, and because you are semipermanently on board, you can study intensively with almost no distraction. As you progress through the ranks, the travel becomes more accessible and your opportunities to see more of the planet expand.

## YOUR LIFE OUTSIDE THE JOB

As noted, cruise work generally is scheduled for three months on and one month off, so this is how your life goes. Your personal life must be orchestrated around this unusual arrangement. It takes some getting used to, but those long months off do feel wonderful. It takes a special person to build a life within these parameters, as well as a special family at home to get used to the schedule. The life is not for everybody, but it is for some. You should consider this venue carefully.

# Professional Organizations

Please see Appendix A for more information on each of the following professional organizations for culinary and foodservice professionals working on cruise ships.

American Culinary Federation

Cruise Lines International Association, Inc.

# Interview

**JOHN DALTON, Vice President of Hotel Operations, Clipper Cruise Line**

**Q / What is your current position and company?**

*A: My current position is Vice President of Hotel Operations for a company called New World Ship Management. I have a partner, and together we run New World Ship Management. We own the ships and lease them back to Clipper Cruise Line. Clipper Cruise Line is a very small cruise line, very different than what most people think of when they think of cruise lines. Clipper has four small ships. We carry no more than 140 passengers. We travel to seven continents. We do not focus on being an amenity cruise ship with the 17 meals a day, gambling, shopping, entertainment, celebrities, and so on. We focus on a much different type of a cruise. Our cruises are mainly educational. They are either historical or they are natural in their content. For example, we do a 14-day historical cruise along the Intracoastal Waterway of the United States that traces the Civil War. On board the ship we have historians, all PhDs and great at communicating their knowledge, talking about the Civil War. The other type of trip we do, and we do more of these types of trips, are natural educational trips. We presently have a ship in Antarctica, one down in New Zealand, one in Belize, and one in Costa Rica. On board these ships are scientists, lecturers, naturalists. We always have a geologist and an ornithologist. And when people see things of natural beauty, these things are explained to them. So when people come off our ship, they have a better understanding of where they have just gone, what they have just seen, and what part it plays in the entire world's ecology. People have fun as well. They really come to Clipper to learn something or see something they can't see for themselves.*

**Q / How many people directly report to you? Whom do you report to?**

*A: I have what I consider a four-tier management. We have a lot of people working on shifts that report to the hotel manager on the ship. The hotel manager on the ship reports to one of our corporate managers. Corporate managers report to me. And those are our tiers or levels. Out in the field we have about 350 people.*

**Q / What does your job entail?**

*A: I have been here 19 years. So in 19 years it really has been a tremendous transition. My responsibilities technically are anything that you would find in a hotel that would be on a ship. I am responsible for food, drink, room service, housekeeping, decorating, refurbishments, maintenance, entertainment, management, marketing, sales. I also do crewing, staffing, hiring, firing, and promoting.*

*So anything you would find in a hotel, that is what I would do. I don't really have anyone I report to directly, but of course my partner and I are ultimately responsible for the company.*

**Q/ What is your educational background?**

*A: I was not terribly academic, and even at a young age my parents recognized that in me, and I recognized that as well. I don't perform well in regular academics. After I graduated high school, I worked in our family businesses while I waited for college, but no traditional school was interested in me. Then my parents read about the Culinary Institute of America. It was called the Harvard of cooking schools, and they thought it was the greatest idea in the world for me because I liked to cook. About six months later, I enrolled in the school. I never really had the desire to cook professionally. I wanted to be in the foodservice industry, and I wanted to understand the bigger picture of the foodservice industry. But I never had any great talent or discipline.*

*So I went to culinary school and saw a bigger industry. Toward the end of school, all my classmates were really intense about where they were going for their internship. It didn't really matter to me because I just knew I was going to be part of the industry. And I saw this little index card that said "cruise line looking for a cook." I called up, and seven or nine days later I was in St. Thomas, working aboard a ship, cooking. My first contract was for seven months, which is a very long time. And there are absolutely no days off. You never get a day off working on a ship. I was 20 years old at the time and young and dumb and willing to do that. I used to get up every morning at 5:30 and get off every night at 10. But coming from a family business, I fully understood the requirements, and I was vested in it.*

**Q/ Do cooks still work those kinds of hours nowadays?**

*A: Well, our contracts are for three months. Our people on board work from 14 to 16 hours a day. That is very normal. But when you work aboard a ship, you work three months on and one month off. In terms of a year, you have three months' vacation. So three months' vacation, if you figure that out and you work it out over a year, it's 90 days a year, almost two days a week. And your time comes in chunks, so you can do whatever you want to do, and I think that has an allure for a lot of people.*

*Living on board is also very cheap. You never need your money unless you go off the ship at night to have a beer or buy a souvenir. You have no bills. We pay for insurance from the first day you work on the ship. We pay for uniforms, too. So sometimes when we first tell people what they are paid on board a cruise ship, they get taken aback a little bit, and they think for that many hours that is not a lot of money. But when they realize that they have absolutely no living expenses, our lowest level of employee on board the ship could easily save or just walk off a ship in a year with $20,000 in the bank. People making $100,000 in Manhattan can't put $20,000 in the bank in a year. So you really have to view it as a great investment.*

**Q /** **So what did you do after graduating from culinary school?**

*A: I spent four years as a hotel manager abroad, and those were the most fun years of my life. I worked on our domestic ships at the time.*

**Q /** **Were you prepared for all the management aspects of that job?**

*A: Probably one of my regrets in life is that I did not do a four-year degree. At the time I was very focused on career, and after I was graduated from culinary school, I just didn't want to spend two more years in school. In retrospect, that was probably a mistake.*

*I certainly suffered in management skills. I made a lot of mistakes that I might not have made had I gone for a four-year degree. I suffered financially. It took me a while to understand how to read financial reports. It took me a while to understand flow of goods. It took me a while to understand the psychology of the employee and employer. I think I would have gotten a little bit more of that had I taken another degree or gone on for an additional degree after graduating from culinary.*

**Q /** **Through the years, what has been your management style? How do you motivate staff?**

*A: Never scream. I have been very blessed with people that I have worked with. I work with some really great people. Whenever I work with somebody new, or when I bring people up the ranks of the company, I tell them the first six months, we are going to be doing everything together. Every letter and memo they send out and every time they counsel somebody, I want to be a part of that. And then once I see that they have the ability to handle it, I give them some authority, and they learn how to use it. I don't believe in forceful authority. There are often people who put out a great product. People either want to be part of this or they don't want to be part of it. It's kind of interesting in my career. I've certainly had to terminate a lot of employees. I would say probably 95% of the people that I terminate send me a Christmas card, and they thank me because it was probably a bad match.*

*Talking about termination, it should never be a surprise. I've never had anybody come into my office when I've had to terminate them and tell me they were surprised. They will always say to me that they kind of knew this was coming. "I know you had expectations for me and they didn't really pan out and I tried but I guess that it just didn't work." And you have to separate that from the person. They are not bad people, just in a bad situation, and it's not the right match. And you try to do it as gracefully as you possibly can.*

**Q /** **What do you love most about your job?**

*A: It's two things that always keep me going. The people keep me going because I work with a great bunch of young, dynamic people. Because of the nature of our industry, we don't get employees staying on our ships for a long time. They may stay two years or three years, and then they move on. That is understandable because they decide to have families and settle down. So I'm always working with new people. And that's fascinating, to be working with young people especially. I like to teach people. I like to see them succeed. I don't mind when they make mistakes, and I tell everybody, you are going to make a lot of mistakes. The only thing you have to be careful about when you make mistakes is that you need to be honest with yourself. You have to own up to it.*

*I also really enjoy being on small ships because there are things we can do that would probably be impossible on a large ship. For example, we had a woman who was gluten-intolerant, so she couldn't have any wheat flour. They had pan pizzas for lunch one day, and the server brought one to her, and she said to her server, "I can't have that." And they said, "Yes you can, because the pastry Chef made it with rice flour." The woman started crying. She hadn't had pizza in 30 years, and she just started crying. She was just so touched that one person would go out of their way to make something special for her. And that is the ability that I have on board these small ships.*

**Q /** **What is the thing you like least about where you are?**

*A: It's a 24-hour business, a seven-days-a-week business. I have ships all over the world every minute of the day, and there are so many things that are out of our control. The industry went through this huge problem with noroviruses—a group of viruses that cause gastroenteritis. It happens all over. It happens on land base, it happens on cruise ships. I actually sit on an advisory board for the Centers for Disease Control on this topic. For cruise ships, it's reportable. On land it's not reportable, so it became this huge issue with cruise ships. You take all the precautions and try to make it not happen. But it does happen occasionally. And when it does, you kind of scratch your head and say, "How can we control this?" I don't have any control over 30-foot seas, hurricanes, and the weather. I don't have control over a ship getting stuck in ice in Antarctica and having to pay someone $150,000 to send boats down to get us out of the ice.*

**Q /** **What kind of advice would you give culinary students starting out now?**

*A: I think it is important to find a school that is commensurate with your needs. Is it strictly cooking you are interested in, or perhaps you are equally interested in foodservice management? You have to know your needs, keeping in mind that this industry is built on serving people.*

*Once you have finished school, it's time to have a game plan, a flexible game plan, because things will always be changing. You have to be very careful about planning your career and make sure you make the best investments. And you can't do any of this without discipline. Discipline is essential.*

**Q /** **What do you see as a trend in the industry?**

*A: The number-one problem in the industry, and we all know it but nobody talks about it, is service. Service in the industry is overall fair. It's been inconsistent, it's untrained, it's undisciplined, it's hit-and-miss. I think the next place for us to go is to make service commensurate with the food that's going out. We go to great restaurants, and we have great food. I think culinary people have to understand that they are involved in a partnership. And the partnership is food and service. It's not just food. You can be the greatest Chef in the world and make the most beautiful food in the world, but if it sits in the window for 15 minutes before a person gets it, the guest loses. I don't think Americans are going to tolerate bad service for much longer.*

## ADVICE FROM A RECENT GRADUATE

Name:       Alexis Kaufmann
Education:  AOS, Culinary Arts, Culinary Institute of America, 2005

***Embrace new ways of looking at and accomplishing goals.***

# Organization

**Figure 6-1**
Sample Cruise Lines
Organizational Chart

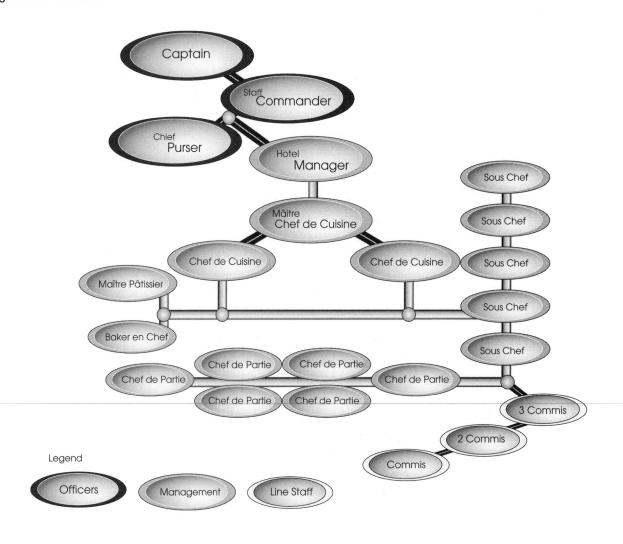

The
Feel   A Day
in the Life   The
Reality   Professional
Organizations   Interview   Organization   **Education Path
Advice**   **99**

# Education Path Advice

**Figure 6-2**
Cruise Lines—
Education
Path Advice

Height of ball indicates importance.

## EXERCISES

1. Compare and contrast the job duties of an Executive Chef on a cruise ship with those of an Executive Chef in a restaurant.

2. Would you like to work as a Chef on a ship? Which activities would you enjoy? What parts of the job might be difficult or not as much fun for you? Could you adjust to the hours?

3. Go the website of one of the major cruise lines (such as Carnival) and find out about the meals they offer.

4. Talk to someone who has gone on a cruise. Ask about the food and meals offered.

5. Using the Career Path Guide (page 2) and Education Path Guide (page 99), map out a possible career and education path for an Executive Chef on a cruise ship, including:
   - Number of months in each job
   - Each place of employment
   - How and when to complete one's formal education
   - How and when to complete skills and competencies

# Part 2

# INTRODUCTION TO FEEDING THE MASSES

The chapters in this part of the book discuss on-site foodservice, which in the past was called noncommercial or institutional foodservice. Typical on-site foodservices include those in universities, schools, hospitals, places of work (such as in business and industry), and the armed forces. Defining on-site foodservice is not easy; usually, it involves food served in a location where foodservice is not the primary business. For example, a hospital cafeteria offers meals to hospital employees who come to the hospital to take care of patients. Likewise, a university dining hall feeds students who are at the university to get an education, and the U.S. Army feeds soldiers who work to protect the nation. On-site foodservices are a part of larger operations with a broader purpose—such as providing health care or education. The foodservice is there as a convenience or benefit, to provide meals to the employees, students, patients, soldiers.

On-site foodservices are different from commercial foodservices in additional ways.

- As an elementary school student, you probably had two choices for lunch: Bring your lunch from home or eat in the school's cafeteria.

You and your classmates were basically what is called a "captive audience" because you had to eat the school lunch if you didn't bring your own lunch from home. You had no other choices. Perhaps as a high-school student you were able to leave the building for lunch to go to a deli or quick-service restaurant nearby. You were not quite as captive as in elementary school, but you probably didn't go out to eat every day, because you had a limited amount of time. The customers in on-site foodservices often have other foodservice choices, but they are not as convenient or are more expensive. You could say that on-site customers are captive to varying degrees.

- Because a given number of people attend a school, work at a bank headquarters, or are hospital patients, you can actually track the participation rate, or what percent of potential customers use the foodservice. Many on-site foodservices, like their commercial counterparts, use marketing techniques to raise customer counts and revenue.

- As a Chef in a commercial restaurant, you have a much smaller group of regular customers than the on-site Chef who sees many of the same people day in and day out. Therefore, one of the challenges of the on-site foodservice is to provide enough variety on a day-to-day basis to satisfy customers.

- Because the on-site Chef knows more about the number of customers who are likely to be coming in for lunch on any given day, he or she can plan purchasing and production with more certainty than can commercial restaurant Chefs. For example, university foodservices sell board plans to their students. Based on the number of board plans they sell, they can have a better idea of their volume of sales, and therefore production volume, than many other foodservices. This is a plus for the Chef.

- Because of the greater predictability of production, on-site foodservices are usually not as frenzied as their commercial counterparts, which experience more fluctuation in customer counts and menu item popularity. Also, working hours tend to be shorter and more predictable in on-site foodservices.

As you can see, on-site foodservices have their own benefits and challenges. On-site foodservices are either self-operated or independent, meaning that the Chef and other employees are employed directly by the institution or managed by an outside company known as a managed services company, foodservice contractor, contract company, or management company. A managed services company is a business that provides foodservices at institutional, governmental, commercial, and industrial locations for a specified period, based on contractual arrangements. The contract may be based on profit and loss, management fee, or a mix of the two.

When a managed services company has a contract with a hospital, for example, to run its foodservice department, the hospital becomes the client of the managed services company. The managers for the hospital foodservice are provided by the managed services company, which may also pay the salaries of the hourly employees, depending on the contract.

Managed services companies provide management to approximately 80% of business and industry, 50% of college and university, and 40% of hospital foodservices. A business may decide to contract out its foodservices to a managed services company for various reasons: to bring in the contractor's employees and save money through reduced payroll, to let the contractor take care of hiring and paying employees, to reduce food costs due to national purchasing contracts, to allow the contractor to renovate or build new foodservice facilities that otherwise wouldn't be developed, and to provide management expertise the client normally wouldn't have. On the other hand, a hospital or other facility may decide to stay independent for other reasons: to retain revenue from cash operations, to eliminate management fees, to allow the foodservice managers to have more flexibility and ability to innovate, to retain control, and to avoid the divided allegiance of managers who serve two masters—the managed service company and the client.

The bottom line is that neither option, self-operated or managed services, is better; rather, each has its niche. As a Chef, keep in mind that managed services companies do generally offer more opportunities to work in varied venues, more possibilities for promotions, and excellent development and training programs. Examples of managed service companies include ARAMARK, Restaurant Associates, Sodexho, Canteen, and the Compass Group.

## FEEDING THE MASSES CAREER PATH GUIDE

Figure P2-1 shows the Feeding the Masses Career Path Guide. It is only a guide for you to get an idea of what type of work experience you need to move forward in your career. The Career Path Guide indicates the number of months in various positions that ideally prepare you for the Executive Chef and other positions noted at the bottom of the chart.

The Career Path Guide should help you understand the time commitment needed to reach top management positions in the foodservice industry. By no means do we claim that these paths are the only way or that the time frames are precise. Some people will excel much more quickly than we indicate, and some may need more time. The length of your workweek, your dedication, and your commitment are all factors in the time frame needed to obtain your final goal. In addition, you may be able to hold two positions at the same time. For example, you could probably get experience as an expediter while working as a grill,

**Figure P2-1**
Feeding the Masses
Career Path

## On the Way
### Nonmanagement positions*

| Position | | | | | | | |
|---|---|---|---|---|---|---|---|
| Prep cook | 8 | 8 | 8 | 8 | 8 | 8 | 8 |
| Pantry cook | 6 | 6 | 6 | 6 | 6 | 6 | 6 |
| Grill cook | 12 | 12 | 12 | 12 | 12 | 12 | 12 |
| Sauté cook | 12 | 12 | 12 | 12 | 12 | 12 | 12 |
| Expediter | 6 | 12 | 6 | 6 | 6 | 6 | 6 |
| Breakfast cook | 6 | 4 | 6 | 6 | 6 | 6 | 6 |
| Banquet cook | 6 | 8 | 12 | 12 | 8 | 12 | 12 |
| Butcher | 4 | 6 | 6 | 6 | 6 | 6 | 6 |
| Roundsman | | 8 | 8 | 8 | 8 | 8 | 8 |
| Retail counter cook | 6 | | 6 | 6 | | | 6 |
| Garde manger | 8 | 8 | 8 | 8 | 8 | 8 | 8 |
| Pastry cook | 8 | 12 | 8 | 8 | 8 | 8 | 8 |
| F. H. service staff | 4 | 6 | 4 | 4 | 6 | 4 | 4 |

### Management positions**

| Position | | | | | | | |
|---|---|---|---|---|---|---|---|
| Sous chef | 12 | 12 | 12 | 12 | 12 | | 12 |
| Banquet chef | 12 | 12 | 12 | 12 | 12 | 12 | 12 |
| Restaurant chef | | 24 | 12 | 12 | 12 | 12 | 12 |
| Executive sous chef | | | 12 | 12 | | 12 | 12 |

**Finish as**

| B&I Exec. Chef Staff Restaurant | B&I Exec. Chef Exec. Dining | B&I Exec. Chef Catering | Regional Chef B&I | Executive Chef Assisted Living | Executive Chef Hospital | Executive Chef University & School |
|---|---|---|---|---|---|---|

*Number of months in the position at 40 hours a week.
**Number of months in the position at 55 hours a week.

sauté, or pantry cook. Similarly, you could simultaneously be a roundsman and a butcher or a garde-manger.

## FEEDING THE MASSES EDUCATION PATH ADVICE

The Education Path Advice guide, found in each of the following chapters, displays the level of importance (through the relative height of the balls) of formal education, knowledge, and competencies needed for positions in different settings.

When you plan your educational path, the key is to remember that education is an investment in the future, on which you will draw over your entire career. You may want to go to a two-year intensive cooking degree program, then work in the kitchen to perfect your skills, and return to school for another two years or so to get a bachelor's degree. Make a plan to maximize your time and your education.

The foodservice, hospitality, and culinary professions are first and foremost hands-on professions, but you should understand that they present minds-on challenges as well. As your hands carry out the precision tasks of common kitchen activities, your mind must retain, compute, and comprehend the important skills you are perfecting. As you absorb the academic aspects of your profession—for example, financial reporting, computer techniques, menu costing, and menu development—the hands-on aspects make these skills come alive. The hands-on experience, which is the passionate and creative part of your profession, complements and motivates all facets of the work. Did you ever notice that when you are committed to a topic, you find that learning about it is easy? You're not studying, you're absorbing; you're not being tested, you're remembering.

Don't be intimidated by the Education Path Advice. Keep in mind that any academic program you need to complete will be filled with examples and situations related to the hospitality field that is your first love. Take one of the subjects in the education path that may sound threatening. For instance, food and beverage (F&B) financials may be scary because of the word financials, which makes it sounds as if the class is all math problems. This is true—but the math problems all represent real-life situations you will need to know how to handle as your career unfolds. Financials and math problems are not as scary when you encounter them in the business you are committed to and passionate about.

Continue to advance your abilities with classes that will be an asset to your career. You could learn Spanish, which is spoken every day in many kitchens. Or you could take a public speaking course to help you do cooking demonstrations or other group presentations. These investments in your future will definitely pay off.

The educational tools we have mapped out are a matrix of the many directions available. We hope you use them to inspire, plan, and evaluate a successful journey toward a rewarding culinary career.

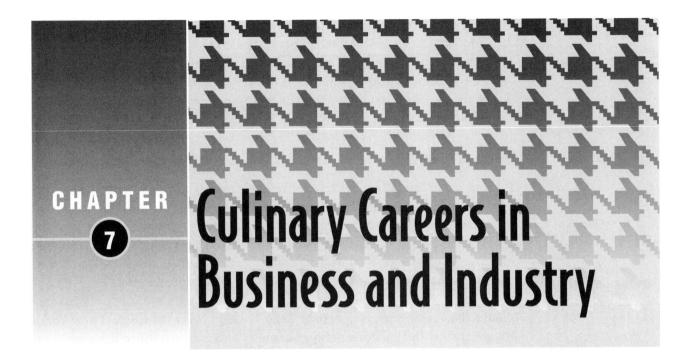

# CHAPTER 7

# Culinary Careers in Business and Industry

## Introduction

**CHEFS IN BUSINESS AND INDUSTRY** prepare meals and snacks for people who are working. Business and industry foodservices (commonly referred to as B&I) provide meals, for example, in financial and insurance companies, manufacturing plants, service companies, and high-tech businesses. Sales in B&I foodservices come from the café (which used to be called the cafeteria, but is often quite upscale nowadays), executive dining rooms, catering, and vending.

B&I foodservices must be innovative to compete effectively with restaurants in the area. Companies also want a pleasant dining room and good food so employees stay on the premises and enjoy themselves during breaks and at mealtimes. A well-run foodservice has a positive effect on employee morale.

B&I foodservices often charge reasonable prices. In some cases, the company subsidizes the cost of the operation. This means that the revenue brought in by foodservices does not completely cover costs, and the company kicks in whatever money is needed to continue operating.

## The Feel

### THE WORLD OF B&I

During my tenure in B&I, I was invited to travel to other facilities to teach, assess, and elevate the culinary level of our company. I was fortunate to visit an array of dining locations. Some were as big as ball fields, others small yet accommodating. The script for B&I properties is huge, from the corporate giants to the small companies sharing centralized cafés. All have a common goal: to feed the business world of America a quality breakfast, lunch, and sometimes dinner. Some run seven-day schedules, but most operate a standard workweek that allows for a normal quality of life.

## THE LAW FIRM

I feel as if I've landed on the set of *Law and Order*. I'm here at one of the most prestigious law firms in Manhattan to evaluate the executive dining room and conference catering. The plush interior and elegant ambiance set the tone for how the company operates both within its walls and externally. My first and most important stop is a fresh latte at the coffee bar. The surroundings are swank and stylish for a cafeteria, better referred to as an employee restaurant.

Breakfast starts with an array of dense and delicious muffins, scrumptious scones, flaky croissants, gooey sticky buns, buttery cheese danish, and New York's finest bagels with all the trimmings. An abundance of fresh fruits in season complements an assortment of yogurt and granola parfaits. There are eggs any style, hickory-smoked bacon, thick maple-cured ham, chicken-apple sausage, Belgian waffles, buttermilk pancakes, corned beef hash, chipped beef on toast, and don't forget the Krispy Kreme doughnuts. Not every facility is this complex, but new standards in corporate cafeterias call for a significant range of choices. This trend will continue to mature in years to come.

Lunch is where it all started, like the first food court found at the Galleria. What is your whim today? Sushi, classic Italian, burgers, pizza or calzone, authentic Indian, or a trendy panini station? There's a salad bar with every topping imaginable and an entrée of the day presented with the care and planning of a home-cooked meal. The beverages, exotic coffees, cappuccinos, and lattes are as creative as at any upscale New York eatery.

## THE BRIGADE

A brigade of culinarians in starched whites lines the perimeter of these food stations, ready to prepare the many choices available to you. Lavish buffet lines with induction chafing dishes are cutting-edge in efficiency and presentation. The look, the style, the feel are important components of the execution of the meals. The pizza station is equipped with an imported wood-fired oven, allowing the smell of pizzeria-quality calzones to permeate the eatery; calzone is today's top choice. The parsley and sun-dried tomato sausage with roasted peppers and onions would give the San Gennaro Feast in Little Italy a run for its money. Every week, many places offer a theme cuisine such as A Taste of Tuscany, Foods of the Mediterranean, Western Hoedown, Texas BBQ, and the foods of many traditional holidays.

In an atmosphere similar to that of country clubs, where you feed mostly the same clientele every day, you need to be predictable even as you surprise your guests with tempting dishes, imaginatively displayed, and mouthwatering demonstrations done to order. There is a balance of well-prepared foods with an emphasis on healthy fat percentage, salt content, and cholesterol count, according to nutritional guidelines. The variety alone can accommodate almost any dietary need. Atkins and South Beach Diet are common terms when lunch is your primary focus.

## THE KITCHEN

The kitchens are contemporary, for the most part, updated and renovated along with the food and seating areas in a time frame acceptable in the industry. Equipment is purchased as trends direct the next culinary spotlight. Tandoori ovens, open-fire grills, and rôtisseries, to name a few items, give versatility to the menu design. The layout of most kitchen facilities compares to that of a hotel, country club, or catering company. The à la carte cooking batteries are professional in design and can satisfy almost any Chef's needs.

## CONFERENCE CATERING

The bread and butter, meat and potatoes of a B&I account is the daily cafeteria, now called a Food Court, Employee Restaurant, Café Dining Facility, or other company-dubbed name. Beyond that portion of the foodservice is a whole world of hotel-style catering, complete with linen, plated meals, banquet staff, and a catering manager. Larger facilities may employ a Banquet Chef dedicated, under the supervision of the Executive Chef, to running last-minute high-level events.

Picture many small conference rooms scattered around a law or accounting firm, investment bank, or other corporate headquarters. These are where client entertaining, board meetings, brainstorming sessions, and weekly staff conferences take place. These functions typically begin with a continental breakfast consisting of sliced and whole fruits, pastries, muffins, bagels with assorted jams, preserves, butter, and cream cheese, and coffee service. The philosophy is that there's no need to go outside when you have the convenience of quality catering a phone button away. For companies, the savings is significant in many ways. Your staff remains on the premises and is therefore more productive, while at the same time your in-house costs are considerably lower than going out.

Catering continues with three-course plated business lunches, in-house meetings with extravagant hot buffets and impressive salads, and all-day workshops with roll-in deli and wrap stations. Creative catering is the tune for these new sources of revenue. Giant office buildings have foyers the size of basketball courts, with atriums three or more stories tall. What a great space for a reception, a plated dinner, a corporate announcement, or the company holiday party. Why? Why not? These questions are being answered by aggressive foodservice companies thinking harder and smarter for their clients and themselves.

# A Day in the Life

### THE FIRST DAY

I wake up every hour on the hour the night before I start my new position. After 20-plus years in the business, why such anxiety? I'm confident, mature, and seasoned in my craft. Why do I keep dreaming of oversleeping? Could it be all the new things I'm going to do tomorrow that I've never done in my life? It's the first time I'm taking a 40-minute train ride to work, the first time employed in New York City, the first time operating a kitchen in a B&I corporate setting—and the first time in 20 years I won't be required to work weekends.

It's 5:30 A.M. The alarm sounds and I jump to attention, like a soldier in the military. The address of the prestigious investment firm where I will be stationed for the next four years is in midtown Manhattan. The world of corporate dining has been a well-kept secret in the foodservice industry. Its reputation from years past was mediocre food, run-down kitchens, and untrained staff. This reputation was fought by the trendsetters in the field who should be recognized for their dedication to converting the way America thinks about corporate cafeterias. The philosophy of corporate dining over the past several decades has changed with the transformation of our eating habits, the demand for higher-quality food, and the need to retain conference catering in-house.

Its 6:45 A.M. when I enter the 30-story glass-walled building, to be greeted by a platoon of security staff. "Morning, Chef," they call as I flash my company ID. "Morning, gentlemen, what's up in the good world?" I reply. "You, Chef, you," they quickly snap back. A thumbs up, and I hop into the private elevator that whisks me up to the VVIP executive dining and conference floor. It looks straight out of an old black-and-white movie: dim lighting, arched ceilings, plush carpets, hand-crafted woodwork, and priceless artwork, not to mention a mysterious corporate ruler who is feared and respected. It's an environment I am used to, given my years of training under demanding and talented Chefs.

There's a buzz on the floor of waitstaff zipping around getting the extravagant continental breakfast ready. The smell of freshly baked croissants, buttery fruit-topped breakfast pastries, and made-from-scratch low-fat muffins are the first scents I experience. The managing directors at this site, my audience, share my belief that breakfast should offer a plethora of choices that are satisfying but not the traditional fat-laden breakfast items. Seventeen of your favorite cereals, all in 16-ounce boxes, are stationed on a gorgeous imported marble credenza for self-service. Note the array of golden pineapples, two kinds of grapes, five whole-fruit selections, and several melons in season, cut into perfectly trimmed wedges. Depending on flavor and time of year, there are ruby red or pink grapefruit, fresh berries, and five kinds of dried fruits, from Turkish apricots to medjool dates. Decorative silver bowls are lined neatly on crushed ice, brimming with freshly cooked egg whites, nonfat yogurts, and low-fat cottage cheese. Ornamental stands line the tables, stacked with New York bagels, country white loaves, Jewish rye bread, and eight-grain health bread, accompanied by a variety of spreads, marmalades, and imported and domestic preserves.

The finale and daily best-sellers are the cooked steel-cut oats and the Swiss-style birchermuesli, complete with organic honey, Devonshire cream, raw brown sugar, golden raisins, and toasted nuts. This fabulous display looks like a page from a *Saveur* or *Bon Appetit* magazine. Not only is the presentation perfect, but the fruits are at their peak of ripeness and the quality extraordinary. Then, just before the directors get tired of the same assortment of items, you, the Chef, dig into your magic file of product tricks and introduce a new twist to your already magnificent breakfast. There is a great deal of room for creativity and freedom at these venues, and in return your clients appreciate your hard work. This is an art, a passion, an obsession that must be preserved and nurtured daily.

As breakfast service commences, the lunch prep is in full swing, with the culinary staff making circular paths around the kitchen like the Indy 500. Intensity, dedication, infatuation, perseverance, enthusiasm are the words to describe the type of commitment this kitchen inspires.

## EXECUTIVE DINING

The air reeks of money. High-energy power lunches, client entertaining, political affairs, financial reviews, and recruiting receptions are all in the palms of your hands. Senior managing directors, superstars, athletes, political figures, and the upper crust are your audience. Menus showcase diversity: from a torchon of foie gras, red onion marmalade, tiny greens, and our own butter crackers, to nutritionally balanced selections, or a Coney Island frank with all the toppings. Let's not forget to mention the killer hamburger and house-cut fries with a fresh-baked dill roll, or appetizers of lobster head cheese, house-made spicy chicken and seafood sausages, jumbo lump crab cakes, and hand-crafted ravioli followed by a selection of fresh fish, certified Angus steaks and chops, lamb three ways, poached chicken breast with keeper squash hash, tiny mushrooms, natural jus, and other composed entrées wrapped up into about 100 covers per day. This is a rewarding experience focused on the pure essence of cooking, flavoring, and technique.

## THE END OF THE DAY

The rush is over, but the staff works diligently to get the place back in order, for tomorrow's another day. It's kitchen downtime, a quiet, peaceful moment when a Chef just sits, stares, and plays back the day's events. Mental notes are made to correct a flaw or to praise a fine job. Critiquing your own work keeps your job fresh and challenging. Tweaking is a word for the lazy; bold changes, risks, and perfection are where my drifting thoughts roam. Well, time to go. Remember! Chefs never sleep; they merely rest. Tomorrow is another day. It's great to never be finished.

# The Reality

In this segment of the industry, everyone's opinion, which includes your client, daily customers, and boss, drives your initiatives. The important thing to remember is that you provide an amenity, a service that is purchased by management and provided for an employee, manager, partner, or guest. If you cease to produce a valuable service (based on the opinion of those you feed), your account will go out to bid to the competition. As every Chef in this business knows, you will need to transfer to another account or find another job. The good part is, if you consistently run your operation with a clear objective—to make sure your customers see the difference you make in their dining world—then you will become a fixture and, barring a corporate change, you will have a good job for years to come. Always remember, your function is to create the repast for a busy executive, a harried secretary, or a particularly demanding partner. You are the oasis in their corporate world, and they will look forward to visiting two or three times a day, if you can keep the oasis green and pleasurable. Use this to your advantage and make their day.

## THE GRIND

To a great extent, you run an operation that requires daily attention and a real understanding of how to supply variety without exceeding cost parameters. Oh yes, you will be given cost guidelines; corporate foodservices often operate at a loss or on a subsidy basis, and you must fully understand and manage the costs as prescribed by the agreement between the foodservice company and its client. All agreements differ in the details. Fundamentally, however, the agreement is that the foodservice company provide the agreed-upon services at a predetermined level of cost, in exchange for a specified fee. With that said, I am sure it is obvious that cost management is a skill in which you must excel.

The fun part is that you must change your menus quite often in order to keep the clientele interested, and you probably will have a reasonable food budget to play with. Of course, your freedom is delineated by context: the type of establishment you serve and the level of expenditure afforded by the employer. Regardless, your challenge is to provide daily offerings of acceptable quality and suitable variety, of perceived value to the employees, in a timely fashion.

The manner in which employees pay for their food varies. They may simply pay for it, or pay but be reimbursed, or receive an allowance. What does not vary is that the diners consider your service a benefit and will be critical of your performance. This is not as threatening as it sounds, for most of us in the industry are driven by passion and love for our cooking style. We take pride in what we do and are constantly challenging our skills and abilities. The main point to understand in this setting is that you feed the same core clientele every day, with a primary client as your liaison.

Although in many ways corporate foodservice is less stressful than a 24/7 operation, you nevertheless must be on your toes, with your culinary bag of tricks close at your side. On the other hand, you may well be supported by the managed services staff, which supplies information for monthly promotions, a repertoire of fabulous recipes, culinary demos, and food shows with trendy new products and equipment.

## PHYSICAL ISSUES

As with most Chef positions, the legs get a good workout, but this life tends to be a bit gentler on the system. Burns, cuts, and bruises are facts of life for all Chefs; however, the activity level is a bit more placid than that in a hotel or restaurant. Your days are shorter, as are your workweeks, but the job is every bit as demanding in other ways. The corporate environment usually features an excellent support system, and your facility is probably in better condition than many restaurant kitchens.

## THE BENEFITS OF YOUR CAREER

Being the Chef in a corporate foodservice is highly rewarding, and both you and your staff will enjoy the workplace. You are the one the clients turn to for a recipe for home or help with a menu for a special occasion. You are the person to whom they turn for dietary advice, cooking tips, and maybe outside catering. With respect to your staff, you are the leader who takes the time daily to teach them, look out for their best interests, and make sure their employment needs are met. In short, you are a very important part of the work lives of many people every day, and that is very rewarding. As with many culinary jobs, there is a family feel to staff relations, and you are the older brother or sister.

## YOUR LIFE OUTSIDE THE JOB

Happily, corporate foodservice work leaves you free at almost human hours to enjoy your life, family, and whatever else you may choose. The job does have built-in late nights and early mornings, but generally you can take off most holidays and weekends. Knowing this, you should of course take advantage and live your life with normalcy. The amount of time you spend on the job usually indicates your level or position in the company. For instance, as your status grows, you will be more involved in opening new operations and guiding the company as a whole.

# Professional Organizations

Please see Appendix A for more information on each of the following professional organizations for culinary and foodservice professionals working in business and industry.

American Culinary Federation

Society for Foodservice Management

# Interview

**BILL CHODAN, Vice President of Culinary Operations, Flik International**

**Q/ What is your current position, and to whom do you report?**

*A: My job is Vice President of Culinary Operations with Flik International. I report to the president of Flik.*

**Q/ How many people do you supervise?**

*A: In my department, there are six corporate Chefs, a nutritionist, and someone who handles the paperwork of our department. So there are eight direct reports, but in theory all the culinary employees of the company and Executive Chefs report to the corporate Chefs and myself.*

**Q/ How many accounts does Flik have?**

*A: We have about 220 accounts in 188 locations. Some accounts we count as more than one. For instance, in a large law firm, we may count the cafeteria as one account and catering and executive dining as two additional accounts.*

**Q/ What's a typical day of work?**

*A: That's a rough one. If there is ever a typical day, it is just trying to return calls and respond to people and respond to problems. Of course, a lot of times I am like a firefighter who gets 20 to 30 voice mails a day. Some of them need immediate attention, such as a Chef resigned or a client wants a special party—things that just need particular attention. Another hat I wear with Flik is that I am in charge of all the account renovations, designs, and construction, so a lot of my time is taken up by meeting with clients and architects about building new facilities, new food programs, and so forth. There really isn't a typical day.*

**Q/ How many hours do you work in a day?**

*A: The minimum is probably 12 hours a day. I leave my house by 6:30 A.M. and get into the office (if I'm going to the office that day) by 7:30. I usually don't leave there or wherever I am until 7:30 to 8:00 P.M., and get home around 9:00 P.M. Very rarely do I get home before 8:30.*

**Q/ How many days a week?**

*A: I would say five is the average. We don't do a lot of weekend work, except when there are special events, we open a new account, or we have renovations in an account. For instance, we are taking over a new account in New Jersey next Friday, so we have to go in there Friday afternoon. Once the old company leaves, we move in and get everything prepped by Monday to open. So the amount of weekend work depends on how many openings and new pieces of business we* take on. *If I'm renovating an account, it's always done on weekends. You go on Friday afternoon, do the demolition work, and then it all gets put back together during the weekend so you can open up on Monday. It depends how much growth we're doing as a company and how many projects are going on.*

## Q / How long have you been with your current employer?

*A: I've been with Flik for 15 years. I started as a corporate Chef with Flik. In its infancy, Flik only had one account in Manhattan. I knew people in foodservice, and Flik was an up-and-coming company. So I called them, and Mr. Flik said that the company was growing and was going to need a corporate Chef. He said if I did a good job for him, that he would eventually make me corporate Chef. So I took the risk, and went to work in this law firm in Manhattan. I did a couple of years there, and proved what I could do and turned the account around. Mr. Flik was true to his word, and made me his corporate Chef about three years later. I was the only corporate Chef, and then the company started to really grow. We started landing law firm after law firm, and we continued to grow. I would open the new accounts, train people, and my role just sort of evolved. Then as the company really grew, we needed more corporate Chefs to assist.*

## Q / What were you doing before you joined Flik? Did you plan out your career?

*A: There was no master plan for me. I wasn't a great student in school; I was probably quite the opposite. I just didn't apply myself. I knew I'd do well at whatever I did, but I didn't know what I wanted to do. I just happened to be cooking in a restaurant when I graduated high school. So over the course of three to four years, I worked in some different restaurants, and then I got an opportunity to go work for a hotel. I thought that would be interesting because at that time you had a lot of different trades in the hotel. You had a Pastry Chef, a Garde-Manger, a Butcher. I thought that could be good, since I had no formal schooling, so I went into the hotel and started as a breakfast cook. I got to work in banquets, with the butcher, with the pastry department, and really learned from some good Chefs and took a little something from each one. Over a course of five to seven years, I started working my way up the hotel to Banquet Chef, Executive Sous Chef, and eventually Executive Chef. I did that for ten years with Stouffer Hotels.*

*At that point, after ten years in the hotels, I knew the only way I could move up was to relocate to other hotels around the country, and I didn't want to get into that. That's when I decided to get out of hotels and look for something else. That's when I found out about Flik.*

## Q / Is there something you would do differently if you could go back now and plan?

*A: I probably would have gotten more formal training because I really had none. It was more just learning from other people. So I may have missed certain elements of a formal training. Actually, I used to go home and practice skills I was weak in. I know that sounds silly, but pastries, for instance—I never really spent a lot of time in the pastry shop, and we had a good Pastry Chef in the hotel, so I would ask him for recipes. Then I would go home and bake something to learn that technique, such as making a great cake or marzipan. But I guess a little more formal training would have helped early on.*

### Q / How did you pick up things like speaking and working well with others?

*A: It was a long, drawn-out process for me. I'm a streetwise kid from the Bronx. I think it was just constant development over the past 20 years. Probably the last ten years with Flik really helped my development. Rudy and Julie Flik taught me that how you present yourself, both physically and verbally, is very important. It was rough going at first. There were actually points in my career, even with Flik, when I felt I might not make it. I didn't express myself as clearly as I wanted, nor was my appearance at the corporate level. But I had good role models who liked me and saw I had skills that just needed to be polished. So, really, it was just a development process over 20 years.*

### Q / What's your management philosophy?

*A: I really think it is treating people fairly and making sure they have fun coming to work. And that's probably another side of me—I try to have a little too much fun at times. If people have fun coming to work, I think they will contribute more. I think it all comes down to respect. You deal with everyone on the same playing field. You can't treat different levels of the kitchen or the hierarchy of the kitchen differently; you treat them the same. You say good morning to everybody, you thank them for coming in every day, thank them when they do a good job. If they don't do a good job, tell them and be honest with them. I think the roughest part of a job is when you have to tell someone they're not doing a good job, and here's why. I can honestly say, in all my years, out of all the people I've mentored, even those I had to terminate, they usually thanked me for treating them fairly. I told them when they made mistakes, told them what to correct, and warned them what would happen if they kept coming in late or whatever the problem was.*

*It's also important for people to like you. When people like you, you'll get more out of them. Let's say you have two individuals and they both have equal skills, but one is a little ornery and sullen at times, comes in with a bad attitude, or is not happy to be here. The other guy is happy to be here, always has a positive attitude, and never says no. If you had to choose between those two people, it's always going to be that person with the better attitude. It has to be sincere; it can't be fake. You can see I want to be here, I like being here, and I'm having fun being here. People have to like you. I would not have moved up as much in this company if people didn't like me and if I wasn't open to change and critiques.*

### Q / What do you like most about your job?

*A: I like being able to control my job. One thing about Flik is that we have always had an entrepreneurial spirit. If we see a great trend out in a little store down in Greenwich Village, I may want to try it out. I can do that and be very*

The
Feel
A Day
in the Life
The
Reality
Professional
Organizations
Interview
Organization
Education Path
Advice

117

*creative. I guess I like the freedom and the variety too. My personality likes to be continually challenged. We are always landing new accounts, new pieces of business, and diverse pieces of business, so I am continually challenged. I prefer not to stagnate in one position, and [don't like] not being able to move out of it.*

**Q/ What do you like least about the job?**

*A: One of the harder things of the job is keeping up with voice mails, cell phone messages, emails—just getting back to all those people and responding to them on a timely basis. Obviously, the best thing is to prioritize them. Even if a vendor calls me and he wants to sell me some bizarre product that I have no interest in, I like to get back to him and say thanks, but no thanks. Because somewhere down the road, you may cross paths again. It's back to respect and treating everybody equally and giving them your time. But it's getting hard to give everybody the amount of time they deserve. At the end of the day, you see you haven't gotten back to a person and feel guilty. I think that's the hardest thing. You have to prioritize your day, and try to get back to everybody.*

**Q/ Speaking to young people who want to end up where you are, what advice would you give them?**

*A: Build a good foundation; really learn all aspects of the business, culinary and otherwise; get good training; understand a little bit of everything in the kitchen. You really have to be a jack of all trades. Make sure people like you, and get their respect. Be trustworthy and dependable. That's probably the two biggest things I would say that's been helpful for me. And try not to lie. Tell people the truth. You'll get caught in the lies. You have to be honest with people, but they'll appreciate it in the long run. Be dependable, too.*

**Q/ What are some of your thoughts about trends in your industry? Where do you think it's going?**

*A: Keeping up on ethnic trends is really important. The great thing here in Manhattan, you can walk around the neighborhoods and you'll see the trends out on the street. I also try to read a lot—I'll read a magazine in 25 or 30 seconds. As I flip through a magazine, if something catches my eye, I cut it out and put it in a folder. Then when I'm challenged or I need new ideas, I reach into that folder. I actually have folders labeled Spring, Summer, Fall, and Winter, and if I see something I like, I'll throw them in those folders. And then, going into the fall, I'll go into the folder and say, "That's great, I saw that in the food show or in the little store in the mall." We do what we call retail tours, where we travel around New York and see some hot shops and new things out there. I just visited a store that sells 40 types of rice pudding with great roasted nut toppings and fruit toppings, and it's all natural. I ask myself if I can take an element of that and apply it to my business. Everybody says there's no new ideas, you just steal the old ones and put your spin on them. I think there is some truth to that. I think it's just rediscovering old cuisines and tweaking them to make them work for you. I think the importance is the authentic use of the ingredients.*

**ADVICE FROM A RECENT GRADUATE**

Name:       Jeremy Moody
Education:   AAS Culinary Arts, San Jacinto College, 2001

*Be a sponge! Take it all in, from the food to the people working around you, because it all plays a part in your future.*

# Organization

**Figure 7-1**

Sample Business and Industry Organizational Chart

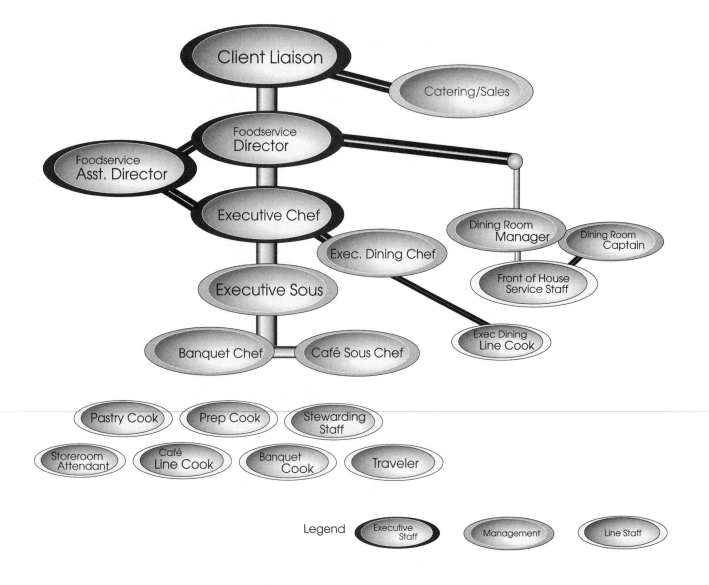

The
Feel      A Day
in the Life      The
Reality      Professional
Organizations      Interview      Organization      **Education Path
Advice**      **119**

# Education Path Advice

**Figure 7-2**

Business and Industry—
Education Path Advice

Height of ball indicates importance.

## EXERCISES

1. Go to the website of Sodexho Marriot, ARAMARK, or Compass. What types of culinary jobs do they have in B&I? Print out a description of at least one job.

2. How would you rate the job outlook for jobs in B&I foodservice: excellent, good, fair, or poor? What does the job outlook depend on?

3. Would you consider working in B&I foodservice? Why, or why not?

4. How is being a Chef in B&I foodservices like being a restaurant Chef? How is it different?

5. Using the Career Path Guide (page 2) and Education Path Guide (page 119), map out a possible career and education path for an executive chef in B&I, including:

   - Number of months in each job
   - Each place of employment
   - How and when you would complete your formal education
   - How and when you would complete skills and competencies

6. Go to the library and find five periodicals that relate to on-site foodservices. List the name of each one, and compare and contrast them.

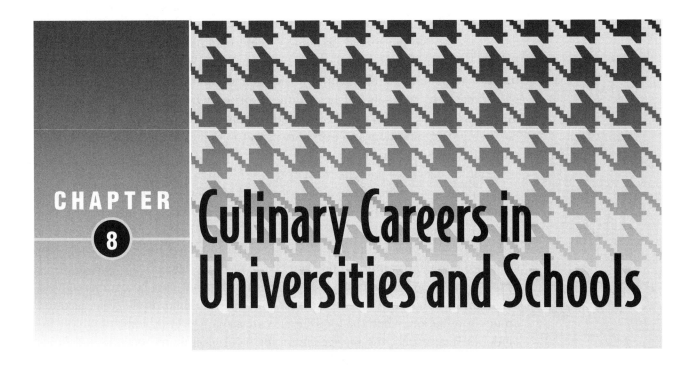

# CHAPTER 8

# Culinary Careers in Universities and Schools

## Introduction

**STUDENTS ON COLLEGE AND UNIVERSITY CAMPUSES** have such sophisticated tastes these days that many creative Chefs have found a home there, cooking comfort foods such as pizza and ethnic foods from around the world. With this impetus, college and university foodservices have changed dramatically from 20 or 30 years ago. Gone are the days of mystery meats in the dining halls and dorms loaded with students who were required to buy full board, meaning three meals a day, seven days a week. Nowadays, less than half of students are likely to live on campus. The rest live off-campus or commute from home. Working as a Chef in the ivory towers is all about keeping residents happy with food selections and quality and working to attract off-campus students, commuters, faculty, and staff employees.

## The Feel

### UNIVERSITIES

University foodservice has three segments that emphasize fresh food, made-to-order food, variety, and authenticity.

- ◆ Residential dining
- ◆ Retail outlets
- ◆ Catering

Residential dining encompasses the dining halls primarily designed for resident students. No more cold breakfasts or unimaginative menus in the dining hall these days! At one college, the dining hall has over ten stations offering traditional menu selections, deli sandwiches and panini, salad bar, soups, pizza, pasta bar, stir-fry and sauté dishes, ethnic favorites, whole fresh fruit, bakery and other desserts, ice cream and frozen yogurt. Some resident students are vegetarian (using dairy products and eggs) or vegan (no dairy or eggs), so both options are available at every meal.

Retail foodservices include restaurants (sit-down and quick service), convenience stores, coffee shops, bakeries, and delivery service. For example, Pennsylvania College of Technology operates eight retail services: a sit-down restaurant, two convenience stores, two snack bars, a café, a coffeehouse, and a pizza delivery service. Retail foodservices blend well-known national brands with signature brands, meaning brands you create or other Chefs have created.

Last, university foodservices do lots of catering that can be compared to a neighboring hotel or conference center. The busiest time of the year is often around graduation, when the foodservice caters many events, some of which are outdoors. Year-round, the service caters for faculty seminars, presidential fundraisers, socials, trustee meetings, and many other events. Catering often requires the foodservice to abandon the normally casual feel and move on to linens, three-course plated dinners, and exquisite cooking marked by superb technique and flavoring. After all, the customers are the powerhouses of academia, the business world, and elsewhere.

The major source of university foodservice income is resident students. Students have a variety of meal plan options when they arrive at college, and the information is encoded on their college identification card. Here are typical options:

**BOARD PLANS** — Board plans are used almost exclusively by students living on campus. Many board plans offer a certain number of meals per week (such as 10 or 14) and flexible spending dollars (often $100). The meals must usually be taken in the residence cafeterias, and the flexible spending dollars can be spent anywhere. Students contract for the number of meals they expect to eat, taking into account factors such as "I never eat breakfast" and "I'm not on campus during weekends." Unused meals are not normally refunded at the end of the semester.

**DECLINING BALANCE PLANS** — With the declining balance plan, residents or commuter students can set up accounts in specified amounts (often $500) to cover purchases. When food is purchased, the purchase amount is deducted from the account. Balances left at the end of the year may or may not be refunded.

**NON-SEMESTER PLANS** — Commuter students may buy a meal plan that includes a given number of meals and flexible spending dollars; the plan lasts until the total number of meals and the flexible dollars are exhausted.

The variety of plans is endless, but one thing is clear: College and university foodservice has a much more predictable volume of sales, and therefore production volume, than many other foodservices. This is a plus for the Chef.

## SCHOOLS

Whereas college and university students will probably love your latest Thai menu creations, don't be surprised when second graders will not eat your kid-friendly vegetables. Children can be picky eaters, which is something the U.S. Department of Agriculture (USDA) acknowledged when they launched Team Nutrition. Team Nutrition is a program that provides training and technical assistance to schools to ensure that personnel have the education, training, and skills necessary to provide healthy meals that appeal to children.

For students in kindergarten through 12th grade, the National School Lunch Program (NSLP) is a federally assisted meal program operating in over 101,000 schools and residential childcare institutions. It provides nutritionally balanced low-cost or free lunches to more than 30 million children each school day. The USDA administers the program at the federal level. In school districts, Chefs often plan menus and develop recipes, and they may supervise food production. Each school district usually decides whether to offer the NSLP and additional programs, and whether to use a contract services company or operate the food-service themselves.

School lunches must meet the recommendations of the Dietary Guidelines for Americans, which recommend that no more than 30% of an individual's calories come from fat and less than 10% from saturated fat. USDA regulations also establish a standard for school lunches to provide one-third of the Recommended Dietary Allowances of protein, vitamin A, vitamin C, iron, calcium, and calories. To provide local foodservice professionals with flexibility, four menu planning approaches yield healthful and appealing meals.

The choice of what specific foods are served and how they are prepared and presented is made by local schools and/or the managed services company they hire. Many school districts involve students in discussing what foods they would like to be served at mealtime. Chefs may develop new recipes and test them with groups of students for acceptance.

As children grow, their tastes start to expand beyond the familiar comfort foods of chicken nuggets, pizza, hot dogs, and hamburgers. Middle schools and high schools often offer à la carte purchases in addition to the school lunch meal. Some schools also get involved in catering and other opportunities to bring in revenue.

## A Day in the Life

Planning menus for students at this state university in upstate New York has certainly forced me to be right up to date on cooking trends and techniques. Most students are more adventurous and know more about ethnic and healthy foods than I remember being when I was in college. Their tastes have become much more sophisticated. Of course, they love the standards: pizza, hoagies, burgers, fries, and so on. But they also love the sushi bar on weekends, foods from Japan or Morocco, and really good vegetarian foods.

On a typical day, I go into work early only if there's a special catering event, such as a breakfast involving the college President. Otherwise, I come in around 8 AM and work through lunch and dinner because that is when our outlets are the busiest. We have two cafeterias, one restaurant, a specialty coffee and bakery shop, and two convenience stores (c-stores) with lots of grab-and-go foods. At night, we have a pizza delivery business, and we also make food baskets for students that we market to their parents.

Every morning, I oversee lunch production and get most of my ordering done. All food production for the cafeterias, catering, and the c-stores is done in one unit—the large cafeteria in which I work. Hot and cold foods are transported to the other cafeteria and c-stores for service. The restaurant uses a different menu and has its own cooking staff, as does the coffee and bakery shop. As colleges have added different venues for eating, the Chef's job of supervision and quality control has grown a bit more difficult. No matter how hard I try, I still haven't figured out how to be in two places at the same time.

During either lunch or dinner, I talk with customers in any of our units. A lot of being a good Chef in a college is keeping customers happy. And what a diverse group of customers we have. Good people skills and just general communication skills are very important.

Afternoons and evenings are often busy catering times, and these are excellent opportunities to do a little public relations with college administration, as they are usually the ones who request the catering. Almost every catered event is unique, and I find them all exciting to design and cook for. I usually work directly with administration on the menu for catered events.

By dinnertime, when I'm getting a little tired, the students are just gearing up to eat dinner, study, party, and more. They love the endless variety of bars we offer, such as pasta bars and sushi bars. They also love having dinner cooked to order in front of them. It's a big challenge to offer new services and programs, such as exhibition cooking, while at the same time restrain expenses, particularly labor expenses. Oh well, every segment has its challenges—but like I said, this job environment keeps me thinking young!

# The Reality

## UNIVERSITIES

A university foodservice, when school is in session, is a 24/7 operation. Food is served morning, afternoon, and night. Catering and special events often take place at night and on weekends. Luckily, though, business goes down tremendously during breaks (holidays, spring, etc.) and during the summer, so you have some time to recharge your batteries, evaluate your programs, and test new recipes and ideas.

Whether you work for a managed services company or self-operate, it is critical to work with staff and the dining services director to increase sales and maintain or reduce costs, especially labor costs. To increase sales, sometimes you get involved in building new outlets or renovating. You must be up to date with the latest in equipment and renovations to successfully attract off-campus students, commuters, and faculty and staff to your foodservice outlets.

In general, food production in universities is varied and challenging from a culinary point of view. To be successful, you need to consult regularly with students to find out what they like about the food, what they don't like, and what they would like to see in the future. Good communication and people skills are essential for dealing with students, working with your staff, and arranging and producing catered events for the President and others.

## SCHOOLS

School foodservice is not as demanding from a time perspective as working in colleges and universities. Only one or two meals are served, and work is often done by the middle of the afternoon. From time to time, events will crop up that require evening hours or part of a weekend. During summer, you are usually off, although that varies. Some Chefs working for managed service companies may be assigned to other accounts for the summer.

As a Chef in a school district, your goals are challenging: increasing student participation, boosting student satisfaction, making healthy meals, having only 30 minutes to get 100 children through the lunch line and eat lunch too, and achieving cost reductions. The profit margin in school foodservice is not nearly as high as in many other types of foodservice, so cost control is vital. With constrained funding for the school lunch program, it might seem easy to just raise prices, but when you do so, participation decreases.

Along with the challenges, you also get the satisfaction of making a difference for the kids you serve, whether they are in first grade or high school seniors. If you enjoy working with children, a career in school foodservice can be fun and satisfying.

# Professional Organizations

Please see Appendix A for more information on each of the following professional organizations for culinary and foodservice professionals working in schools and universities.

American Culinary Federation

School Nutrition Association

National Association of College & University Food Services (NACUFS)

# Interview

**DONALD R. MILLER, CEC, CCE, AAC, Executive Chef, University of Notre Dame**

**Q / What is your current position at the University of Notre Dame?**

*A: I'm currently the Executive Chef of the Morris Inn, a hotel and conference center, which is a $4 million food revenue a year operation. But in the next couple of weeks, I'm going to be promoted to the Executive Chef of Notre Dame foodservices. The food revenues are around $40 million.*

**Q / How many meals for students are there on an average day?**

*A: Just for the students, it's 28,000 meals a day. And that's actually relatively small. What people don't understand is the University of Notre Dame has only 12,000 students, not including graduate students.*

**Q / And how many venues are there for them?**

*A: We have two student dining halls: North and South. You go into South Dining Hall and it is set up in an international format, everything from Mongolian wok to pasta. We have American, Italian, South American, and other cuisines. With their meal plans, students can get their meals at North or South dining halls, or go into any of our satellite operations.*

**Q / How many satellite operations are there?**

*A: Quite a few. In the Student Center is the Huddle Food Court, which includes branded concepts such as Starbucks and Burger King. We have three restaurants on campus and seven express eateries, each with a different theme and menu. All are open to the students, faculty, staff, and public.*

**Q / How is the foodservice in the hotel organized?**

*A: At the hotel, there is an Executive Chef and an Executive Sous Chef. There is a Pastry Chef and then, of course, there is the structure of AM and PM Sous Chef and then all the line staffing cooks, like a normal hotel operation would be structured. The hotel makes 95% of the menu items from scratch. The kitchen has been renovated and now includes a refrigerated butcher room which we also use for Garde Manger work, pastry specialties, and cold hors d'oeuvre presentations. Now, in the college foodservices, there is a structure of an Executive Chef and then what we call Unit Chefs, which would be much like Sous Chefs. But in some of the satellite outlets they don't necessarily need the culinary training that one would need at the hotel level or our satellite restaurants, just cooks and pantry Garde Manger levels. Universitywide, we employ in the area of 165 to 175 cooks.*

**Q / Who does all the baking for the foodservices?**

*A: We have a centralized operation that includes a bakery, butcher shop, and a vegetable processing kitchen. It also includes the central warehouse for all food products. It's a tremendous, fascinating facility that makes all their own breads, rolls, danishes, as well as thousands and thousands of cookies. And there is an area where they do cake decorating, finer à la carte desserts, chocolate work, and the more creative pastry workmanship.*

**Q/ So what is a typical day like for the Executive Chef of Notre Dame's dining services?**

*A: Nothing in this job is typical, but let me tour you on what I perform as a daily routine. The Executive Chef is responsible for the culinary integrity, unit operations, culinary operations, and their costs. Once a month there is a mandatory meeting of the culinary staff leads, all the Unit Chefs, Sous Chefs, Pastry Chef, and of course the Executive Chef. We talk about everything that is going on within our units so we can create a synergy in our areas. The problems are brought up, and everyone can help each other out. So that would be something that would be a regular weekly routine along with the department management meetings, which would include all the satellite operations.*

*On a daily basis, I report in the morning to the office, which is above South Dining Hall, and read the daily reports, check and answer email, look at the daily costs. From then on it's a situation where 30 to 35% of the time I will be out at the different units, 20 to 25% of the time I am involved in training, and then the rest of the time I will be working with menu development. And you know, of course, that there's always more meetings to attend.*

**Q/ Let's step back a little bit and talk about how and why you got into this profession.**

*A: People have asked me that question before, and I think this career picked me. But if you go back further to when I was a kid, I was influenced by my grandmother, who came here from Austria. She was like the cook of an estate in Austria and came here before World War II. In her big home in Kenosha, Michigan, she would be doing all the cooking, and I would do things like crush the black walnuts to go into her traditional Austrian bread.*

*While I was in high school, I worked as a dishwasher in a restaurant. I also got to prepare salads, shred cheese, and perform other simple food preparation jobs. Then, one day one of the cooks didn't show up, and the Chef asked me if I would help work on the line. That was fun and exciting. I worked there for three years when I moved with the Chef in his position at a new Holiday Inn. I worked at the opening, which was round-the-clock setup, as the Banquet Chef. I was really in over my head. The food wasn't unsafe, I just wasn't doing basic cooking techniques—I was young and clueless.*

*After high school, I applied to go to the Culinary Institute of America and got in, but I was put on a waiting list until a spot came open, which was close to two years. In the meantime I went to the University of Wisconsin for two years. In retrospect, I wish I had graduated from Wisconsin and then gone on to Chef school after that. Also, when I dropped out of the University of Wisconsin to go to culinary school, my parents did not talk to me for about a year.*

*The two years that I was at the University of Wisconsin, I took a lot of business classes, and I fall back on a lot of those classes, especially the systems that they taught me. And of course, it's not good to have a system unless you have the discipline to implement it and follow up on it.*

*At culinary school, the first semester was so hard for me, I almost flunked out. Unfortunately I went there with the attitude that I'd been cooking seven years, so I knew it all. I struggled a lot because I went in there thinking it was going to be easy. But I was wrong. Cooking is a science and a discipline that needs to have a foundation of basics. By the last three semesters of school, everything finally clicked for me, and I got it.*

**Q / So after your culinary training, where did you go to work?**

*A: I worked in a huge Holiday Inn hotel at Holiday Inn's corporate headquarters in Memphis. I worked for them for a year, then I found out Opryland Hotel was opening in Nashville, Tennessee. It was an exciting place — over 2,000 rooms and lots of restaurants. So I had an interview there. I walked through that place and it was incredible. Absolutely incredible. So I took the job. I was the Second Saucier in the gourmet dining room, called Old Hickory, which is all copper service. We all worked 17-hour days to get the kitchen in shape; it was incredible. I was Second Saucier, and the First Saucier was Peter Fousch, who was on the 1968 Olympic team. Now all of a sudden I'm being introduced to a whole level of focus that was a fabulous culinary experience. It was like an awakening for me. I was so excited, all I could think about was food and cooking. I was working with Peter Fousch, who taught me how to set up the station as he displayed his European discipline within his own station. We were doing 450 people a night, copper service, that was taken out and plated tableside.*

*I trained for months as the First Saucier and took it over, and then later was promoted to Sous Chef, which was an honor because they normally filled that position with Europeans because of their training. I worked there for another year and then a resort, called Sandestin, opened up in Florida, and they needed an Executive Chef to open up the hotel. I got the job, and it was my first Executive Chef's job ever. I stayed with that company for several years before I found I was burned out from the long hours and tiring work.*

**Q / How do you define "burned out"? Why did it happen?**

*A: I was very passionate, but I had nothing left to reach into. I had never rested—I worked every day of the week. So here is how I got out of it. I applied for a job teaching—one opened in Joliet Junior College. I thought I'd never get it because I didn't finish my bachelor's degree, but I got the job and taught for five years, which got tenure. Teaching helped my burnout and helped me resolidify my base so I could go out again into an industry position.*

**Q / And then from there you went to the University of Notre Dame?**

*A: Yes, there was something I liked about the academic atmosphere of being at that junior college. But here I was, I think I was 33 years old at the time, and I felt like I wanted to be back. I missed the life. It's in your veins. But I wanted that academic background. I looked, waited, and watched, and when this job came up at Notre Dame, I came here, interviewed, and I've been here ever since.*

**Q/ Now tell me a little bit about your management philosophy.**

*A: I'm open to constructive ideas. I like taking those ideas and working on them independently and also working on them together as a group. I like challenging myself and others, and I like other people challenging me. With a great rapport, you can raise the caliber of the kitchen to amazing levels.*

**Q/ As you talk about nurturing your staff and really developing them, is there an apprentice program here that is very intense?**

*A: We have a wonderful apprentice program here. There are four apprentices at the university who rotate through the hotel, all the foodservice outlets on campus, which gives them a well-rounded experience. They go to school at our local community college to gain the academic classes required by the American Culinary Federation. We have strict guidelines and demand the best from our students, as they receive the best from us. In the apprentice program here at the University of Notre Dame, you start in stewarding. You go through every aspect of sanitation, cleaning, and equipment maintenance needed to run a kitchen. The next step is to become a breakfast cook and go through that entire segment, rotating through all the necessary stations. This continues through all stations until they have completed their 6,000 hours of kitchen training. This program is sanctioned by the ACF apprenticeship program, which awards them a Certified Cook certificate upon completion. The program is also recognized of course by the U.S. Department of Labor.*

**Q/ What do you love most about your job?**

*A: I'm passionate about my work. I've heard this before, that the culinary field is the only profession in which you have to use and develop all your senses. You have to educate all your senses. This is a hands-on profession where you also use academic abilities. For instance, I have to crunch numbers to analyze financial progress. If you were to tell me that I need to keep food cost at 31% next year, I know how to position my kitchen structure to achieve that goal. And yet at the same time I've developed my taste buds, my sense of smell, as well as all the dexterity that comes with physical skills.*

**Q/ What advice would you have for a young culinarian who is choosing this profession?**

*A: When you come out of school going into this industry and have your mind set to become an Executive Chef, you have to pace yourself first of all. You have to be patient, put in your time, and build your foundation by developing your technical skills and your personal repertoire. I'm not suggesting by any stretch of the imagination that you do what I did, because working every day was what we had to do back then to prove ourselves to the European market. Keeping in mind dedication, commitment, and perseverance are still your number-one priorities.*

**Q/ What do you see as trends in the industry?**

*A: I feel that the industry is going toward fresh, wholesome, well-balanced (nutritionally) foods, not adulterated by poor cooking skills. The customer demands a perceived value that is not based on huge portions. It is based on well-prepared foods, fairly priced, with a new focus on presentation. Our clients are looking for a relaxed atmosphere and great service complemented by creative, well-prepared foods. This is what I have based my career on, and it has worked for me.*

## ADVICE FROM A RECENT GRADUATE

Name:         Gabriel Gomez
Education:   AAS Culinary Arts, Houston Community College, 2006

> ***Stay focused and constantly create and strive for goals so you grow and your career takes off.***

## Organization

**Figure 8-1**
Sample University
Organizational Chart

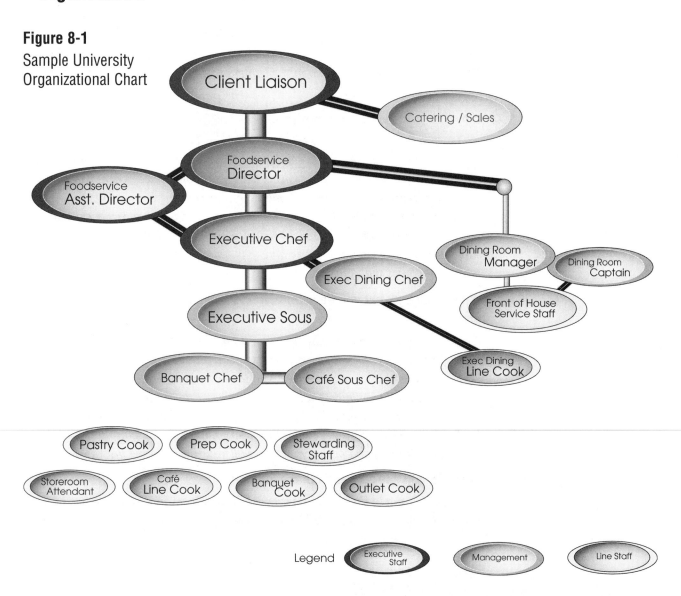

# Education Path Advice

**Figure 8-2**

Schools and Universities—
Education Path Advice

Height of ball indicates importance.

## EXERCISES

1. Would you consider working in universities or schools? Why or why not? If you would, what type of university or school would you like to work in and where?

2. Look at the job listings for university and school jobs at websites of managed services companies such as Sodexho-Marriott and ARAMARK. Print one out that sounds exciting to you.

3. Using the Internet, find a school lunch menu and a university menu. Compare and contrast the menus from a culinary point of view.

4. If you are enrolled in a college, find out about what types of catering the dining services provides.

5. Interview someone who works in a school or university foodservice. Find out what he or she likes about the job and the challenges.

6. Go to the NACUFS website and list the benefits of becoming a member.

7. Using the Career Path Guide (page 2) and Education Path Guide (page 131), map out a possible career and education path for an Executive Chef in a university foodservice, including:
   - number of months in each job
   - each place of employment
   - how and when you would complete your formal education
   - how and when you would complete skills and competencies

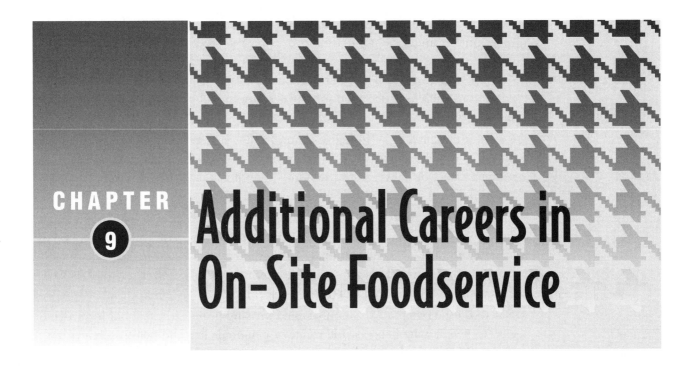

**CHAPTER**

**9**

# Additional Careers in On-Site Foodservice

## Introduction

**ON-SITE FOODSERVICES** also provide meals for patients and employees in health-care institutions such as hospitals and retirement communities, for the members of our armed forces across the United States and around the world, and for people on planes, in sports stadiums, and in prisons. This chapter will discuss the varied settings for culinary professionals in the broad field of on-site foodservice.

## Healthcare Foodservice

Healthcare foodservice has its own excitement, challenges, and avenues for creativity. Imagine preparing food every day in a hospital for:

◆ 250 hospital patients who use a room service menu to order meals
◆ Close to 1,000 hospital employees looking for meals and snacks in the Café
◆ Dozens of catering affairs that routinely include dinner service for the hospital President, Board of Directors, and donors.

Hospital food is changing, and foodservice directors are constantly looking outside healthcare to hotels, business and industry, and colleges for new ideas. Hospital foodservice sales are growing too, and employment benefits are great, including many weekends and holidays off for chefs and managers.

The healthcare field also encompasses long-term care, an area that is becoming more important as the population over 65 years old continues to increase. About one in every eight Americans is over 65, and this group will more than double over the next 25 years. The elderly frequently have disabilities or chronic medical conditions such as high blood pressure, so there will be lots of job opportunities in this area. Long-term care choices include:

COMMUNITY SERVICES — In many communities, services and programs are available to help seniors. These programs often include meals available at senior centers or delivered to the home (like Meals-on-Wheels). Over 3 million Americans receive these types of services every year.

ASSISTED-LIVING FACILITIES — These are facilities in which residents often live in their own room or apartment within a building or group of buildings and take some or all of their meals together. Although they can be quite costly, assisted-living facilities provide older or disabled adults with the help they need as well as social and recreational activities.

NURSING FACILITIES — Commonly called nursing homes, these facilities provide care to people who need a wide range of personal care and health services, as well as 24-hour skilled nursing care.

CONTINUING CARE RETIREMENT COMMUNITIES (CCRC) — These communities, also called life care communities, extend different levels of care based on the resident's needs. The same community may include individual homes or apartments for residents who still live on their own, an assisted-living facility for people who need some help with daily care, and a nursing home for people who require higher levels of care. Residents move from one level of care to another based on their needs but remain in the community. CCRCs generally charge a large payment before you move in (called an entry fee) and then charge monthly fees. Many guarantee lifetime shelter and care.

## HOSPITALS

Working as a Chef in a hospital foodservice is a satisfying job. You are helping the patients get better by making great food that meets their nutritional requirements. Most of the patients get a regular diet, but some need less sodium, less refined sugar, and so on. No, you don't interact with patients every day like the nurses do, but many Chefs visit patients frequently. Just like in a restaurant, you need to get the input and feedback of your customers. Of course, patients are a very different kind of customer because they don't want to be in the hospital; they would much rather be home. In any case, I was a nervous wreck the first day I went to visit some patients with the dietitian. Of course, I survived, and I am much more comfortable now on the floors with my Chef's hat on, doing public relations with the patients and the nurses.

In a hospital, the foodservice is often called the food and nutrition services department. The reason for this name is simple: The department not only prepares and serves food but also provides nutritional care and education to patients. Food and nutrition services is one of the support services of a hospital. Other support services are housekeeping, laundry, and maintenance. The Chef normally reports to the Director of Food and Nutrition Services, who in turn reports to a hospital administrator (usually the administrator in charge of support services).

The organization of the food and nutrition services department varies depending on size and function, but it often includes a Director and two or more Assistant Directors (Figure 9-1). One of the Assistant Directors is usually the Nutrition Manager, or Chief Dietitian, who oversees the nutritional care activities, and the other is the Chef or Executive Chef. The Chef in a hospital foodservice has not just one but several groups of customers: patients, hospital employees, and hospital visitors.

Before going any further, you need to know a few things about the environment in which hospitals operate. Government agencies and insurance companies reimburse hospitals for many expenditures, so the healthcare environment is rather regulated. With pressure from government and insurance companies to reduce escalating costs, hospitals have worked for many years to cut costs and increase revenue in new ways. In addition, the federal government and other entities pressure hospitals to provide quality services and ensure patient satisfaction with those services. The hospital Chef knows and understands these trends and takes actions to reduce costs and increase revenue and satisfaction.

### Patient Service

Hospitals vary in how patients order meals, how the food is prepared, and how the food is assembled and delivered. Service is not as simple as in most restaurants. Patients order meals using a paper menu, spoken menu, or room service menu. Traditionally, patients receive paper menus on their breakfast trays. The menu contains the food selections for the next day and, after the patient fills it out, is picked up and returned to the kitchen. The menu is usually a one-week cycle menu. Because the length of a patient's stay in most hospitals is less than one week, the one-week cycle suffices for variety and is easier and less expensive than a two-week cycle. This method of sending up menus for the next day on the breakfast tray sounds simple but has its share of problems: The menu may not be picked up; the patient may not realize they are selecting for the next day; the patient may not have a pencil or may have other difficulties completing the menu; the patient may change his or her mind, and so on.

With the spoken menu system, a host or hostess visits the patient just before meals to review the menu options for that day. The patient can ask questions and make selections. The host or hostess then returns to the kitchen, assembles the meals, and delivers them.

Room service menus are similar to those in hotels, except that patients are encouraged to call for meals during certain time periods and the service is normally not available throughout the night. Meals are usually delivered within 30 to 60 minutes of ordering. If the patient is not capable of calling down to place a meal order, a nurse or family member may do so.

In most hospitals, food is prepared just prior to service. The amount of scratch cooking and convenience cooking varies from hospital to hospital. In an effort to save money, some facilities use cook-chill systems (a food-production system in which large quantities of foods are cooked, chilled, and then reheated) or even get cooked foods from central commissaries.

Assembly of the individual meals is often centralized in the main kitchen or may take place in smaller kitchens closer to the patient floors. In hospitals where patients fill out their menus a day ahead of time, trays are commonly assembled on a trayline. Trays, dishes, and all foods are positioned along a conveyor belt, and employees are stationed to place specific items on each tray as the trays flow past. The first employee on the line usually puts the tray mat, silverware, and completed menu on the tray. As the tray starts down the line, hot and cold foods are added, and the last person checks the tray contents against the menu to see if it is correct. All food items are covered, and the trays are stacked into carts or trucks and taken to the hospital floors, where they are delivered directly into each patient's room.

If the hospital uses the spoken menu, each host or hostess assembles the trays of the patients from whom they have taken menu orders. Stations are set up in the kitchen to make it easy for the host or hostess to assemble the meals quickly. For example, he or she may start at a condiment station; continue to cold units to pick up juices, salads, desserts, and milk; and pick up hot food from the cooks and coffee last.

As you can see, the newer systems involving the spoken menu and room service are closer to the services available in hotels and restaurants. Patients have been quite satisfied with these options.

### Café

With fewer inpatients and more outpatients, hospital foodservices have jumped into the Café business, as well as other nonpatient services, to increase departmental revenues. In the old-fashioned hospital cafeteria, the food was boring, the atmosphere was stale, and the décor was unappealing. Nowadays, it is common for Chefs and directors to take a retail approach, meaning that they use restaurant strategies to target and compete against a specific group of concepts such as fast food, independent, or chain restaurant. By serving more upscale food items different from what customers expect in a hospital cafeteria, Chefs have created a retail setting and built perceived value.

ARAMARK, which has more than 1,200 healthcare clients nationwide, calls their hospital cafeterias by new names: cafés, on-site restaurants, or retail foodservices. The cafés often resemble food courts, fast-casual restaurants, or high-end buffets with display cooking stations and recognized brands. At the 5151 Café at Zale Lipshy Hospital in Texas, the independently run foodservice changes its self-serve bars, such as Asian Noodle Bar, Greek Sandwich Bar, and the Pasta Bar, every day. Most cafés use a cycle menu, to ensure lots of variety.

## Catering

Catering is an integral part of most hospital foodservices, especially in medium and large facilities. The activities hosted within hospitals—meetings, seminars, receptions, and parties—often call for catered services. Chefs develop standard menus for most catering but create one-use menus when needed.

## Additional Retail Services

Hospital foodservices have branched into many other retail areas such as the following:

CONVENIENCE STORES — For example, Zale Lipshy Hospital has a store off the hospital lobby called the 5151 Market. They bake cookies, pop popcorn, and serve other grab-and-go foods such as soups, salads, sandwiches, and to-go plates of the daily specials from the 5151 Café. Sundries like toothpaste and other personal products are also available.

KIOSKS OR CART SERVICE — As you enter York Hospital in Pennsylvania, you will immediately see (and smell) a Starbucks cart offering a variety of drinks, baked goods, and sandwiches. One employee staffs the cart, which can be moved to other locations as needed.

HOME MEAL REPLACEMENT — With so many hospital employees running home after work to make dinner, some Chefs have capitalized on this by offering dinners to go. For example, Zale Lipshy Hospital offers what they call Friday Night Feasts. Their specialty is gourmet cooking for two. Employees simply order, pick up their meals from the Nutrition Services office, and follow the heating directions at home for a delicious meal. Most items can be reheated with a microwave.

BAKERIES — Another convenience to offer employees in a hospital is baked goods such as birthday cakes, cookies for Mother's Day, decorated cakes for Halloween, and so on.

BRANDED FAST-FOOD RESTAURANTS — Some hospitals, typically larger ones, have contracts with fast-food restaurants to operate units on their campuses. For example, on the first floor of Children's Hospital of Philadelphia is both a café and a McDonald's.

VENDED FOODS — Vending is yet another retail possibility to feed employees and visitors. Vending is often used to feed the night shift in hospitals.

## Dietitians

In the hospital setting, Chefs work with dietitians to develop patient menus and recipes for modified diets. The over 50,000 registered dietitians (RDs) in the United States constitute the largest and most visible group of professionals in the nutrition field. Individuals with the RD credential have specialized education in human anatomy and physiology, chemistry, medical nutrition therapy, foods and food science, the behavioral sciences, and foodservice management.

Registered dietitians must complete at least a bachelor's degree from an accredited college or university, a program of college-level dietetics courses accredited by the Commission on Accreditation for Dietetics Education (CADE), a supervised practice experience, and a qualifying examination. Continuing education is required to maintain RD status. Most RDs are members of the American Dietetic Association and many are licensed or certified by the state in which they work. In addition to working in hospitals, they also work in other healthcare settings, private practice, sales, marketing, research, government, restaurants, fitness, and food companies.

In the hospital setting, you may also work with registered dietetic technicians (DTR). To become a DTR, one must complete an associate degree and curriculum requirements of a program accredited by CADE, then take the registration examination for dietetic technicians to become credentialed. DTRs often work assisting registered dietitians or in supervisory positions in the food and nutrition department.

## CONTINUING CARE RETIREMENT COMMUNITIES (CCRC)

### Levels of Care

CCRCs are unique in that they provide various levels of care within one community for elderly residents whose needs change over time. This is referred to as a continuum of care. There are three stages of care:

INDEPENDENT LIVING — Virtually all CCRCs offer independent living. The main purpose of this unit is to provide a sense of independent living for older adults who are capable of doing the basic chores of everyday life but who may need occasional help. Units come in a variety of forms including studio apartments and one-bedroom, two-bedroom, and even single homes such as cottages. Services such as meals, housekeeping, and laundry service are usually available.

ASSISTED LIVING — Available in over 80% of CCRCs, assisted living is an intermediate step between independent living and nursing care. It provides assistance for residents with chronic care needs that do not require 24-hour skilled nursing care. Assisted-living services include helping a resident with bathing, dressing, taking medications, and other daily activities.

NURSING FACILITY (NURSING HOME) — Almost all CCRCs offer skilled nursing care in a nursing home setting. Most nursing homes offer round-the-clock nursing care for those who require it. Care is provided under the supervision of licensed nurses and other medical professionals such as physical therapists under the supervision of a licensed physician.

Some CCRCs also have special units such as those for residents with Alzheimer's disease.

CCRCs combine a variety of services to improve the wellness of residents: good meals and nutrition, exercise, social activities, and educational activities. Independent living residents are usually offered dinner every evening in a dining room setting with waitstaff and may receive other meals as well. Different communities offer different styles of service. For instance, a CCRC in a major city may serve meals in a classy restaurant-style setting to cater to residents who are used to city life. A community in the Amish areas of Pennsylvania, on the other hand, might serve family-style meals where the food is passed around the table as if you were at home. Distinctions like these make an enormous difference in the overall atmosphere of a community.

The kitchen prepares and serves three meals a day to residents in the assisted-living unit and nursing home. Most of these residents also eat in a dining room setting.

Menus for independent and assisted-living residents are often restaurant-style, with lots of daily specials to add needed variety. Buffets and food bars such as salad bars are popular, as they offer lots of choices and self-service.

# Armed Forces

Every day, over one million meals are prepared in military kitchens. Some kitchens prepare thousands of meals at one time, while others prepare food for small groups of people. Culinary careers are possible in all of the Armed Forces.

- U.S. Army — Today's Army is a highly trained team. Soldiers build bridges, calibrate and operate computers, and apply state-of-the-art tools and methods to solve critical problems. They operate tanks, fly helicopters, and launch missiles. Working together, they enable the Army to accomplish its mission to deter war, but they are prepared to fight and win when war is not deterred. This is important to know because, as a military Chef, you must have this training as well.
- U.S. Navy — The Navy operates throughout the world to preserve peace. Navy cruisers, destroyers, frigates, submarines, aircraft carriers, and support ships are ready to maintain the freedom of the seas. Navy sea and air power are available to assist in the defense of allies or engage enemy forces in the event of war.
- U.S. Air Force — The men and women of the Air Force fly, maintain, and support the world's most technically advanced aerospace vehicles, including long-range bombers, supersonic fighters, and many others. These forces are used whenever and wherever necessary to protect the interest of the United States and its allies.
- U.S. Marine Corps — The Marines are a part of the Department of the Navy and operate in close cooperation with U.S. naval forces at sea. The Marine Corps' mission is unique among the services. Marines serve on U.S. Navy ships, protect naval bases, guard U.S. embassies, and provide a quick, ever-ready strike force to protect U.S. interests anywhere in the world. All Marines can move on short notice to match up with equipment stored on floating bases on the world's oceans.

◆ U.S. Coast Guard — The Coast Guard regularly performs many functions vital to maritime safety. The Coast Guard's most visible job is saving lives and property in and around American waters. The Coast Guard also enforces customs and fishing laws, protects marine wildlife, fights pollution in lakes and along the coastline, and conducts the International Ice Patrol. The Coast Guard also monitors traffic in major harbors, keeps shipping lanes open on ice-bound lakes, and maintains lighthouses and other navigation aids.

Whether you are on a base, ship, or submarine, your culinary efforts will be appreciated by the personnel you cook for. After all, your skills are key to keep the morale of the troops high. In addition to cooking for troops, you may be selected to prepare gourmet meals for formal dignitaries and distinguished guests, both within and outside of the United States.

In the regular service, personnel serve on a full-time basis. After enlisting in a service, they are sent to basic training. Upon graduation, they are sent to specialty job training schools, after which they are assigned to a station for duty. After 20 years of regular service, they qualify for a military retirement.

The reserves are part-time military soldiers. Personnel serve an initial period on active duty after attending basic training and job training. After the training period, which usually lasts several months, reservists are free to return to civilian life, but for the remainder of their service obligation they attend training sessions and perform work in their job specialty one or two days a month with their local unit. Once a year, they participate in an active-duty training session for 14 days. When reservists have completed 20 years of service and have reached age 60, they are entitled to retirement based on reserve pay.

The military distinguishes between enlisted and officer careers. Enlisted personnel begin at the lowest rank in the military and serve as the main workforce, making up about 85% of the Armed Forces. With education and service, advancement can be expected. Officers begin at a supervisory rank and are the leaders of the military, supervising and managing activities. They must have a four-year college degree from an accredited institution before being commissioned. The military has several programs that lead to becoming a commissioned officer, such as officer candidate schools, Reserve Officer Training Corps (ROTC), and direct appointments.

Some service members enter the military as enlisted and convert to the Warrant Officer Corps. The Warrant Officer Corps is made up of specialized technicians with solid leadership experience who serve an entire career as a Warrant Officer in their chosen field. Warrant officers are also commissioned and receive the same privileges as senior ranking enlisted and commissioned officers of all ranks. The Army and the Marine Corps both have Warrant Officers in the foodservice career field.

In order to enlist, one must be between 17 and 35 years old, be a U.S. citizen or an alien holding permanent resident status, not have a felony record, and possess a birth certificate (some requirements vary depending on the service). Applicants must pass both a written examination and meet certain minimum physical standards, such as height, weight, vision, and overall health. All branches require high school graduation or its equivalent for certain enlistment options. Most active-duty programs have first-term enlistments of four years, although there are some two-, three-, and six-year programs. The contract also states options such as sign-on bonuses and the types of training to be received.

Foodservice specialists in all branches of the military are enlisted personnel who prepare all types of food according to standard recipes and modified recipes for special diets. They also order and inspect food supplies and keep kitchens and dining halls clean. Foodservice managers are officers who perform some or all of the following duties:

♦ Manage the cooking and serving of food at mess halls.
♦ Direct the operation of officers' dining halls.
♦ Determine staff and equipment needed for dining halls, kitchen, and commissaries.
♦ Set standards for food storage and preparation.
♦ Estimate food budgets.
♦ Maintain nutritional and sanitary standards at foodservice facilities.

These are just a few examples of military culinary jobs.

## Additional On-Site Foodservices

On-site foodservice also includes in-flight foodservice, recreational foodservices, and correctional foodservices, which are discussed here.

In-flight foodservice provides meals and snacks to people traveling on airlines. As you can imagine, in-flight foodservice is quite specialized because the food must be prepared, stored, transported, and then, in the case of hot foods, reheated on the plane and served. Tens of thousands of meals are prepared and delivered to planes for in-flight service. Scheduling is particularly important to avoid delaying flights.

Airlines often outsource foodservice to catering companies, such as LSG Sky Chefs, that specialize in preparing meals for flights. Caterers are awarded contracts, much like a managed service contract, and prepare the food in commissary kitchens located near airports. Serving freshly made food is a challenge when it may be hours before the food is served on the plane.

Although some airlines are cutting back or even eliminating in-flight food, there is no clear trend. Other airlines are putting more emphasis on good food and hiring world-class Chefs to plan menus and oversee food production. Chefs who work in this segment may become members of the International Flight Services Association (see further information on IFSA in Appendix A).

Recreational foodservices include food served at sports areas and stadiums, theme parks, racetracks, state and national parks, zoos and aquariums, and even the Olympic Games. Some recreational foodservices are run by managed services companies. For example, ARAMARK has much experience providing foodservices for the Olympics. Food choices offered in recreational venues have improved over the years. For example, baseball stadiums have always been known for their hot dogs and popcorn, but these days more ballparks are offering local dishes and upscale foods. For example, San Francisco's 3Com Park has vegetarian burritos and a 40-clove garlic chicken sandwich on the menu. In the Northeast, cheese-steaks are a best-seller in Philadelphia stadiums, and clam chowder is popular at Boston's Fenway Park. In some cases, a national brand such as McDonald's operates in a zoo or other recreational facility.

Correctional foodservices provide meals to people in prisons and jails. Correctional foodservice managers provide nutritious, cost-efficient meal service for confined populations. The American Correctional Food Service Association (ACFSA) is an international professional association serving the needs and interests of foodservice personnel in the correctional environment. ACFSA sponsors conferences that help its members sharpen management skills, promote professionalism, and keep up-to-date on developments in foodservice and corrections.

## ADVICE FROM A RECENT GRADUATE

Name:       Robert Geosits
Education:  Culinary Arts, Sullivan University, 2007

> *Respect those who have been in the business and learn from them and their mistakes.*

## EXERCISES

1. Find a hospital and a continuing care retirement community in your town, city, or state. What are their names, and where are they located? How big is each facility? Do the facilities have websites? What do the websites say about food or nutrition services?

2. Go to the website of St. Paul and Zale Lipshy University Hospitals, www .zluh.org. Briefly describe their retail foodservice concepts.

3. Go to the website of Sodexho-Marriot, Aramark, or Morrison Healthcare Food Service. What types of culinary jobs do they have in hospitals or CCRCs? Print out a description of at least one job.

4. Would you consider working in healthcare foodservice? Why or why not?

5. What different responsibilities would a hospital Executive Chef have from an Executive Chef in a CCRC?

6. Visit the websites of American Society for Healthcare Food Service Administrators and the National Society for Healthcare Food

Service Management. What is different about the memberships of these organizations? Which organization seems to offer more resources for its members?

7. Would you consider working in armed forces either full time or part time? Why or why not? If so, which branch would be of interest?

8. Talk to someone who has been in the military. Find out which service branch they were in and for how long. Ask them what they liked and didn't like about their service.

9. What benefits does a military job offer that a civilian job doesn't offer? What obligations does a military job have that a civilian job doesn't have?

# INTRODUCTION TO HOME ON AND OFF THE RANGE

In Part 3 we discuss nontraditional careers. In other words, these Chefs are not usually cooking in a kitchen for guests in the dining room. Instead, they might be doing any of the following.

- Developing new products for restaurants, retail, and manufacturing.

- Teaching the culinary arts in high schools, postsecondary school, or cooking schools for nonculinary professionals.

- Planning and cooking meals in a private home.

- Purchasing all food and beverages for one or more foodservice establishments.

- Styling or photographing food for newspapers, magazines, cookbooks, advertisements, movies, television shows, or theater productions.

- Writing about food and cooking for a newspaper, magazine, or other publication, or authoring cookbooks.

- Promoting a foodservice operation, food company, or other company as a public relations manager.

In Chapter 10, you will read about the exciting world of research and development chefs, also known as Food Innovation Chefs or Product Development Chefs. Chapter 11 discusses culinary educators, and Chapter 12 delves into the other careers noted here. Nontraditional careers for Chefs are available more now than ever before. Depending on your interests and skills, one or more of these careers is sure to interest you.

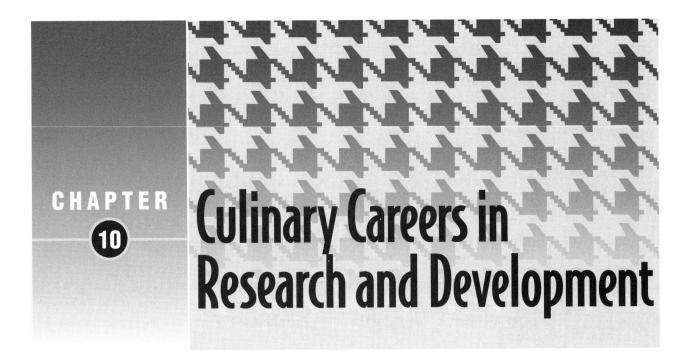

# CHAPTER 10

# Culinary Careers in Research and Development

## Introduction

**RESEARCH AND DEVELOPMENT (R&D)** Chefs are getting a lot of attention these days, and understandably so. These hybrids, who are part Chef and part food scientist, use their cooking and marketing know-how to develop and evaluate new menu items for chain restaurants, food manufacturers, multi-unit operations and other companies. They blend their culinary training with education and experience in food science. All chefs develop creative recipes that focus on taste and are visually appealing, usually served right from their kitchen or from an off-premise site. The quintessential difference in R&D recipe development is not only the recipe focus, but R&D chefs must have a broad scope of what is called good manufacturing practices. Good manufacturing practices are the general procedures that ensure both quality and safety of the food product. These disciplines include a detailed understanding of food preservation, mass production, and Hazard Analysis and Critical Control Point (HACCP) food safety and sanitation, as well as fluency in the technical terms and products used by food scientists. R&D is vital in all facets of food service, from product development, flavor profiling, FDA label regulations, and requirements for manufacturing food products that may be sold in many outlets across the country. Additionally they analyze and reproduce flavors as facilitated by the available technology. Most large food processing companies have test kitchens staffed with food technicians, research assistants, food scientists, and professional Chefs. These R&D chefs are mixing good taste with good science, which equals great products. The process of creating new menu items involves a great deal of precise planning and work. To develop prototype menu items, or enhance existing offerings, research Chefs look broadly at marketing research, customer data and input, food trends in restaurants of all types as well as the media (television, cookbooks, magazines), information on ingredients, and much more. Once a prototype is developed, research Chefs will perform an example of the following.

- ◆ Determine recipe cost.
- ◆ Help to determine what price can be charged.
- ◆ Analyze the nutritional content.
- ◆ Work with vendors to develop specifications.
- ◆ Develop equipment, methodology, and presentation of item.
- ◆ Create the recipe, analyze and make replicable in any location.
- ◆ Present the product to tasting groups to gather feedback and insight.
- ◆ Test the item with real customers.

Multi-unit restaurants then test promising new items in selected markets before determining whether to roll them out throughout the company. Many prototype menu items fail to get to this step. Thus begins the culinary carousel of R&D.

# The Feel

So who are these research Chefs, and where do they work? Most research Chefs have worked in the culinary field for more than 20 years. The research Chef needs strong culinary knowledge and skills and extensive restaurant experience. Operational experience is important so the Chef understands menus, customers, and how new menu items are integrated into a restaurant. Because so much of the job involves communication, the Chef also needs excellent oral, presentation, and writing skills. Research Chefs work in many settings: all types of restaurants, food manufacturing companies, hotels, cruise ships, healthcare, schools and universities, business and industry, ingredient supply houses, and even flavor and spice manufacturers. The culinary talents of Chefs are becoming more involved in research and development in part because consumers have a more sophisticated palate and demand foods that taste great in a variety of food establishments. Supermarkets, gourmet food stores, chain and multi-unit restaurants and manufacturers over the past decade have embraced the culinary community to raise the quality and meet their clients' expectations. The seamless partnership between Chef and scientist has been revolutionary in creating numerous career opportunities in the culinary profession. According to a survey conducted by the Research Chefs Association, research chefs consistently earn toward the top of the chef pay scale—on average, from $70,000 and up.

Research Chefs are often members of the National Restaurant Association, American Culinary Federation, and the Research Chefs Association (see Appendix for more information). The Research Chefs Association (RCA) was founded in 1996 and now has over 2,000 members, including research Chefs, food scientists, and other food professionals in sales, marketing, manufacturing, distribution, and the media. As you can read on their website (www.culinology .org), RCA members are "the pioneers of the discipline of Culinology™—the blending of culinary arts and the science of food." RCA is helping colleges develop a culinology bachelor's degree programs that offer coursework in culinary arts, food science, business management, nutrition, processing technology, and government regulations. The amount of colleges and universities offering food science programs well suited to pursue a career in R&D is continuing to grow and expand their curriculums.

In the following pages of this chapter we have captured the essence of a career in the field of R&D. For decades this profession has had little exposure as a dynamic culinary choice, sometimes being portrayed as having little hands on cooking and creativity. We hope through our researching these opportunities you discover the misconceptions of this new frontier filled with well respected and talented professionals.

# A Day in the Life

### COFFEE FIRST:

Whether tradition or not, there is something comforting about that first cup of morning coffee. I call it "my time" when all others have to wait as I make my daily plan of attack, anticipating the tidal wave of emails, prospects, deadlines, new projects and prototypes that will swallow up my day. The real world of research and development is not as foreign to the rest of the food service industry as many believe. All chefs everywhere formulate recipes, develop procedures, calculate quantities, spec out ingredients and create presentations of their recipes. This has been going on as long as civilization has been eating multi-ingredient based dishes. The classic difference in R&D is the recipe formulation is just one component of the final product. A single-unit independent chef creates a recipe, formulates the method of preparation, prepares and presents the final product. The whole process can take as little as a few hours. The R&D process usually starts with an idea or expectation from your customer. Once the concept is outlined you will spend hours of questioning, brainstorming, troubleshooting and evaluating before any preliminary testing and recipe formulation begins. Usually the R&D team will solicit client feedback before any physical preparation commences.

### THE CLIENT—WE ALL HAVE THEM IN SOME WAY, SHAPE OR FORM

"I want the moistest, most delicious decadent chocolate cake ever, for a great price, beautifully boxed, packed and shipped to 15 central warehouses." Or you may hear this. "I have this fabulous, rich, sinful white chocolate cheesecake. Can you reproduce the exact cake, make it freezer stable for 6 months, in batches of 2500 lbs, portioned, baked and glazed in 1-lb 6oz cakes, individually boxed, palletized and shipped around the country?. . . . Oh, I forgot to mention for half the amount it currently costs us." Remember without clients we have no business, so listen, evaluate and digest before you begin to comment. The variety of products is endless, from desserts, dips, side dishes, sauces to entrees, hors d' oeuvres, snacks and soups. Many products that you see in a multitude of eating places are all part of the R&D process. With the demand for higher quality prepared foods, the career opportunities in this area of food service will grow with the need for qualified R&D chefs.

The client scenario will initiate the detailed process of formulating, or in many cases tweaking a recipe, to fit within the guidelines necessary for its application. Every product that goes into development has many answers to questions before any testing can begin. You can spend weeks working to perfect one recipe, with multiple challenges, so your ability to overcome failure must be composed, while still remaining confident until you are able to achieve success. Let me walk you through a snapshot of how a typical product in R&D might start.

A new product may either be an existing recipe that a customer wants duplicated, or a new concept clearly mapped out with objectives established by current industry trends, or the expectation of today's consumers. If the product doesn't exist, the role of the chef is to develop a product from "scratch" using the data and information of what is popular or where the needs of the consumer are headed. The preliminary research analysis of the target market combined with the recipe development including cooking method, cooking time, scale-up process and final roll-out can take a team weeks and even months to finalize. It is essential for an R&D team to be proactive and remain intuitive to future trends because the process of product research from start to finish can be very arduous. Development in this sector is a collaborative effort. The cost analysis of the recipe is scrutinized by finance for profitability. The marketing team gives their approval on popularity, while food scientists verify, among other things, the ingredients, sustainability, food safety and other production obstacles. The operations team confirms that a restaurant crew will be able to produce the same quality in their own environment.

In R&D it is our responsibility to create options for our clients. We find inspiration by following trends in market research, eating at fine dining restaurants, reading culinary magazines and recognized cookbooks, or by traveling aboard. Once the fine tuning of a recipe is devised with the criteria clearly defined, tested and perfected in the lab, the product is ready for critiquing of the consumer and client.

## TESTING—HUNDREDS NEVER LEAVE THE LAB

There are hundreds of alternative creations that never make it out of the laboratory, given the many different variables products take into consideration. A recipe that may appear to be a great fit will often not make the cut once it goes through the rigorous testing of R&D. An item for consumption will need to have increased shelf life without altering the integrity of its taste or texture. The product may be produced in one location, then transported to another (such as a distribution center), then many times shipped to a third location for its actual sale. In many instances it will be cooked or reheated by a novice. Another criteria would be the longevity of the product, being particularly concerned with safety and handling. Keep in mind when a chef creates a recipe for 50 or 100 portions the scaling of the ingredients in not proportionally affected by the increase. In a product development situation, you are usually scaling ingredients to batches of 2000 lbs or more to perhaps decrease cooking time or perhaps save money. If the product is to be frozen or must become more shelf stable, there is a need to test how well it holds up using different stabilizers for frozen, refrigerator or shelf sustainability.

My first instinct when I begin to analyze, is to turn the product inside out. Every question should be asked to insure better productivity leaving less room for error. Questions such as "Do I grill or pan sear? What vegetable cut makes the most sense? Should we use fresh or frozen vegetables for this application? What level of spiciness will be better received or what is the best cheese to give it the ultimate taste and texture profile? Keep in mind the product must maintain a standard of convenience, as well as possess quality and consistency needed to attain a place in the market.

These are just some of the steps and scenarios that take place so a product can be successful in the hands of the end user. When the exercise of testing a product is complete, including scrutinizing and reformulating a recipe, gaining rave review under all the criteria for its application in the market, there is no better accomplishment. I used to cook for hundreds and thousands of people who complimented and applauded my food. Now I reach millions! I get such satisfaction knowing I am making a difference in how Americans are eating today and what are they putting in their bodies. I create wholesome, many times organic, all natural ingredients into products for human consumption, that's what it's all about. At the end of a project it's equally important to congratulate your crew and the other supporting teams. As we all know, a great leader is nothing without the team. My trance of reflecting gets interrupted by the (back to business) ring of the kitchen phone. I pick up the phone with a confident, kind of arrogant tone, "Hello chef." I wait and listen. "Yes Sir, sure we can do that, no, not a problem!! We'll get on that immediately". Everyone looks at me when I hang up as if to say, "Could you wait a few minutes?" but they know, you're only as good as your last meal. And in a flash the group breaks into their designated places and we are all going back to the drawing board. The accolades of our last success quickly dissipate, but the pride hovers over us as we gather to discuss yet another new adventure in the arena of research and development.

## MY CAREER

I started my career in the "traditional" kitchen environment, long hours, fast pace, intense service of restaurants and hotels. I have always loved the creative aspect of the industry, watching, learning, absorbing any and every opportunity to master a new dish, perfect a new method or take a new position. I received my AOS degree in Culinary Arts, figuring at a young age, my education is done, I'm on my way to be an executive chef, I was good to go. Months turned into years and with the blink of an eye, 10 years of dedication, commitment and hard work went flying by. I loved the restaurant rush, tiring nights of serving and executing large events or small VVIP's, but at the same time reality set in and I felt it was time to get out of the day-to-day operation of high-powered kitchens. I felt there were numerous other career opportunities in our profession, but I was somewhat limited due to my education. With much apprehension and enthusiasm I registered for school in pursuit of my bachelor's degree in hotel management/ business. I was worried about the academic requirement of school, while I continued employment to support myself. A typical situation for someone going back to school at this stage of their career. Despite my need to work, I was confident I made the right decision and determined to finish my degree.

## MORE SCHOOL?

As we all know the education of "Culinary" has mostly the element of hands-on learning and on the job mentoring. I quickly realized as I continued my studies that there were components of academics that I would need if I was to further my path in Hotel Restaurant management. Finance, marketing, business, strategic planning and management were some of the courses that would be critical for my success. I attended the University of Nevada, via online courses and some physical attendance, transferring 30 credits from my previous AOS degree. The whole process, due to the limited amounts of credits I could handle in conjunction with my continued employment, took me three and one-half years to graduate and receive my bachelor's degree in Hotel Restaurant Management. Again, I felt I hit another huge milestone in my life. I was ready to focus my career on the management side of the profession, or how we chefs like to call front-of-the-house or "FOH".

I found employment through the help of the college's placement department, in a Grand Casino on the strip of Las Vegas. I was ready and willing to make the move to front house management, but when I interviewed for a mid-level Food and Beverage manager position, the Human Resources director was thrilled I had such a strong culinary background. She immediately called the Executive Sous Chef to meet me. She hung up with the Chef and turned to me saying, "This would be a great place for you to start to get to know the property and help straighten up our Grand Cafe." I didn't mind being pulled back into the kitchen short term, but my goals now were to stay in food service on the management side. So there I was with a BA in Hotel Management working back-of-the-house (BOH) yet again. I knew right then I wanted more.

My quest to pursue my MBA was very important, or this scenario of struggling to get my first management gig after 13 years in BOH was going to follow me through the rest of my career. I began my task to continue with my MBA at Michigan State

University. While I was enrolled in their program, I was reintroduced by some of my professors to the updated curriculum of food science. The business of R&D was a new wide open market with detailed focus on value, stability and quality. "Pre-fab" foods that are tasty, well produced and affordable were becoming a necessity in the American way of life. Television in the mid to late 90's was creating a more sophisticated client, generating awareness of the cooking profession that never existed before. We were teaching Americans how to cook with an assortment of ingredients, showing proper methods and techniques to create restaurant quality foods. The only problem is that no one in their busy schedules has time to cook," from scratch". So either you as a family are out 4-5 nights per week or you are picking something up at the store that requires minimal cooking expertise, but demanding dining out results. The question being raised was who was going to produce this standard of cuisine that would be found in chain restaurants, supermarkets, and food market or gourmet shops? In order to stay up with this trend, food companies in every area were starting to include culinary experts in all facets including recipe development, and testing of new products and menus. The explosion of natural, organic, and wholesome ingredients were appearing on food labels, backed by manufacturers that believed in quality preparation. They began replacing unhealthy foods filled with fats, sugars and preservatives, with the necessity of science blended with the proficiency of the chef.

The idea of R&D as an alternative profession began to fascinate me. I wanted to be part of this industry revolution, knowing with my years of professional kitchen experience I could bring a plethora of information. But, this is the beginning of a new initiative in food service. I had to ask myself, were there really enough jobs in R&D that would incorporate my knowledge as a culinarian with my education and understanding in food science? The answer I found from market studies, peer mentoring, and college, university placement centers proved this exploding giant of R&D could foster huge opportunities for multiple job levels and increased future advancement.

## "PH"-WHAT D?

I was hooked with the idea of mixing culinary creativity (my passion), combined with the science of food, (the mystery). What I never would have believed if you had approached me 13 years prior, was that I would be enrolled in Michigan State University getting my Doctorate degree in Food Science. I would have simply told you as a young aspiring culinarian, "Sorry pal, wrong guy". I was enamored with the opportunity to continue in this educational path. Remember this was in the late 90's when the world of R&D was just starting to embrace culinary professionals as viable members of their universe. There was food science, and there was culinary creativity. The incorporation of the two was in their infantile stages. The RCA, Research Chefs Association, was just formed, spreading their insight and commitment of Culinology®, the universally recognized integration of culinary arts with food science. They were visionaries, lending their support of courageous culinary professionals pursuing a career in R&D. Many other schools offered a variety of food science programs and certificates classes, but I felt if I was to gain respect in this newly formed area, I would need the highest level of education. So once again after 13 years in the Food Service business I continued my quest in food science education. There I was hitting the books and by the way, loving every minute of it.

During the time I was studying for my PhD, a transformation began to take place that is hard to explain. The missing links of the many scientific phenomenons of cooking and chemical reaction of foods began to make sense. How does emulsification happen? Why does lemon or lime juice cook raw products? How does caramelization take place? The understanding of cooking methodology and traditional flavor profiling I replicated for years, but never could explain its mysterious results, was now becoming crystal clear. The science and the art became picture perfect in my evaluation of how to incorporate the two. I was fortunate to clearly appreciate the fine art of classic cooking, and now I could add the component of manufacturing from a plant pilot to a scaled-up retail shelf. This knowledge gap in my understanding of food service has opened up a whole new world, blending nutrition, science and culinary artistry, into appealing, high-margin products.

I finished my PhD in Food Science in 1998, from Michigan State University. With my history in the food service/ hospitality industry and my ability to promote and network myself, I was fortunate from that moment to hold a number of prestigious positions surrounding R&D including international & national food chains, mega markets and food manufacturers. I became involved in several professional organizations that are comprised of chefs and food scientists working together to face the many challenges in R&D. I stayed committed through these organizations and educational institutes to utilize my expertise to influence the research and development industry. Given my years of hands-on experience in R&D, I am confident I have helped others with the many obstacles they face, offering solutions or alternatives to their many issues.

## THE R&D KITCHEN

My kitchen is similar to any professional food service kitchen. Heavy-duty stoves, salamanders, ovens, table top mixers, steam and convection ovens, stainless tables lined with color coated cutting boards and state-of-the-art knives, vegetable processors, under counter refrigeration, tile floors, marlight walls and even walk-in coolers and freezers. The main differences are that our measurement tools are designed for exact precision. Graduated cylinders stand in for traditional measuring cups, scientific balances that measure to the milligram replace standard electric or baker's scales. R&D chefs or food innovators use the same technical skills to produce foods that will taste great in large batches under real professional kitchen conditions. The exemplary difference is how we use various extracts and flavoring compounds that, although natural and many time organic, take meticulously accurate handling. These zesty granules and liquid taste sensations can reproduce the essence of almost any known ingredient with the expertise of a food scientist/flavor specialist.

# The Reality

Being a research Chef has its share of benefits. It allows more time for creativity because the job does not entail running an every day F&B operation, dealing with customers in your territory on a daily basis, employees calling out, or a constant concern about food and labor costs. The hours are longer than 9 to 5, but they allow for more flexibility than many other businesses. The salary is financially competitive. The salaries are at the top end and beyond what an excellent chef would make. The development of any item is a shared effort with many other departments. This equates to meetings that will highlight your non food skills and test your ability to persuade and prove your hunches on most new projects. You must be able to articulate your ideas and relate them to a group successfully, especially if they require a large monetary investment.

You are the boss in your area, but keep in mind your whole existence is dictated by consumer trends, customer's lifestyles or by a client with a specific expectation. Your experimental recipes are at the mercy of focus groups every step of the way. They may pick items from a book of pictures with descriptions of menu items or you might prepare a tasting of 20 or so items depending on the situation. The suggestions that are from customer input are for the most part very general requests like, large portion size, lower pricing, certain flavor profile such as smoky or sweet. These customer groups may or may not like what you have created, which sends you back to yet another planning stage. Keep in mind you are not cooking for yourself, you are cooking for what the consumer wants to buy. Your job is to collect the data from these groups and make the necessary changes needed to make these winning products. Yes, your expertise and advice is essential in this process, but at the end of the day, the final decision is out of your hands.

## THE GRIND

We would all be lying if we told you there was nothing monotonous about your daily activities. As a culinarian, one of most difficult tasks is to sit in one place for a long period of time. This professional has its share of strategic meetings, planning sessions and critiquing round tables that can test your patience and tolerance for the marketing component of your position. The only word of advice is, the better prepared you are going into these situations, the shorter the amount of time you will spend evaluating and reiterating an issue. When you are in developmental stages of a product, the redundancy of recipe evaluation or tweaking the elements associated with a product can be a tedious and boring exercise. When that happens, remember the bigger picture. Everything in R&D must be perfect or you are jeopardizing the success of the final product. Not always does your creation take a boring turn during its development, but it happens and must be noted. Lastly when you think of a chef, you think of a mad scientist, secretly mixing a variety of ingredients, tasting, then adding a pinch of this and a splash of that with no regard for accuracy or replication. Remember in R&D every splash, dash, and pinch must be painfully measured, documented, costed and replicated taking it from the R&D bench to the retail shelf.

## PHYSICAL ISSUES

Don't think you are getting away from the physical part of the culinary profession. Granted the quantity of your ingredients may be less then the 100 pound sacks of flour or 50 pound bags of carrots we have all flung over our shoulders. But walking, agility to glide gracefully around your work station quickly, bending, standing in one position for a length of time is again standard for anyone when you are talking about working in a kitchen. You will chop, dice, butcher, pound and peel your share of ingredients, keeping in mind the principles of food handling and safety will achieve a new level. Furthermore you will spend tiring hours sitting at a computer, creating the other half of the intricate puzzle to complete your final masterpiece. Additional exercise in this field, or any other parts of our industry should be a given to keep you from medical issues later on.

## THE BENEFITS FOR YOUR CAREER

The benefits of working as a research chef extend far beyond financial compensation. Most restaurant and hotel chefs usually work weekend, holidays and nights because that is when your clients are off and want to enjoy dining out or vacationing. The R&D chef usually works within business hours. Even with the deadlines, client demands and projects an R&D kitchen run at a more relaxed pace unlike the quick tempo of a large volume restaurant. The chance to be creative and innovative adds additional spice and passion to the job. I love having the ability to fine-tune a recipe that will be enjoyed by 10's of thousands of people, sometimes millions. In the R&D business your primary focus is to anticipate culinary movement, gather consumer information, identifying what they crave now and where they are headed in the future. In this field you are always months ahead of the surrounding curve, placing you in a position to set the next trend.

## YOUR LIFE OUTSIDE THE JOB

Your world feels like one big open book. Your next new discovery could be an ethnic favorite at a family picnic, or a simple trip to the local food market, or even a city street fair, where worldwide dishes and local specialties are the flair of the day. This industry allows you more time to get involved in activities outside your daily job responsibilities. You have the ability to attend national trade shows, symposiums and continuing education, giving you the aptitude to stay up to date on what's new in R&D, fostering future career advancements. You have the luxury to make advanced plans and commitments usually without last minute change due to an on-job disaster, which is conducive to a more normal lifestyle.

# Professional Organizations

Please see Appendix A for more information on each of the following professional organizations for culinary and foodservice professionals working in restaurants.

American Culinary Federation

Research Chefs Association

# Interviews

### LORI DANIEL, Chief Inspiration Officer/Chef, Two Chefs on a Roll

**Q / What is your current position?**

*A: My current position is Chief Inspiration Officer/Chef, and my role is new business development, marketing, and ambassador of culture for my company.*

**Q / And your company is?**

*A: The name of the company is Two Chefs on a Roll, Inc., and we are a custom food manufacturer. I cofounded the company in 1985 with Eliot Swartz, my husband. We create products for retail, food service, and manufacturing. The product range is from dips, sauces, pestos, puddings, to a full range of desserts. Customers are unique retailers or the top-tier supermarkets such as Trader Joe's, and restaurant chains such as California Pizza Kitchen. For manufacturing clients, we co-pack recipes that they supply to us, or we will create a specific product for their needs. Currently we co-pack several labor-intensive food items for different companies. Often the opportunities that come our way are products that they're not geared up to do, such as a new product launch for which the equipment is a significant investment or something that is extremely labor-intensive, particularly on the dessert side. On the dessert side, we are very hands-on. On the savory side, we're highly automated. We are at a turning point right now in our company, trying to decide what our new identity is going to be.*

**Q / How big is Two Chefs on a Roll?**

*A: We have 230 hands-on employees and a full office support staff. We relocated our facility from 33,000 to 100,000 square feet, building the plant, negotiating the lease, getting the building done, the contracting. It was a major undertaking, and then moving into the new plant with a new overhead, a new breakeven, new responsibilities, so obviously we want to grow it much larger. We've been working on putting the organization together and bringing on a Vice President of Operations and a Chief Financial Officer. We even have a Chief Information Officer. For a food manufacturer that is our size, we have quite a team, and that's what we need to be in place to continue to grow.*

*The challenge is we only have so many resources. We have some bank loans and we have an SBA [Small Business Administration] loan, but we haven't been able to get any more money than what Eliot and I put against it. So, if we put in 250, they put in 250. It does inhibit your growth, and it kind of chokes your resources a little bit. And now that we're spread over those three markets—retail, food service, manufacturing—it's hard to market the company in those three arenas. Trade publications are different. Buyers are different. Products are different. People are all different.*

**Q / What was your career path up until Two Chefs on a Roll?**

*A: How did I choose the food industry? Well, for me it was sort of by default in some ways. I wasn't a very good student, and I found that I loved to work and did well. So, when I became a senior in high school, there were these work-related occupational programs that you get out of school at 11 and go to work. It made all the sense in the world to me. One of the work programs was called home economics–related occupations. Well, I figured I could probably get into that one, so I signed up for it. I had to find a job in a kitchen even though I didn't know how to cook. I found a job as a prep cook at the nicest steak and lobster house in town, but only after convincing the interviewer that I could lift 50-pound bags of onions, put up with the heat, and so on. They told me that "we don't hire girls," but somehow or other I got a phone call from them to start work. I fell in love with the work. I loved being on my feet. I loved the excitement. I had been involved in drama in high school. So this was like being in theater.*

*I was 16 years old, and I just knew I had found a place for myself that combined a lot of the things that I liked. But understanding a career in it, I had no clue. But part of this program was you had to spend time in the college resource center. I knew with my grades I was never getting into college. So here I was just flipping through brochures. I didn't even know what the word culinary was, and I fell into it. I was looking under cooking, I guess, because I didn't know what else to do. There was this brochure for the Culinary Institute of America, and I had to look up the word culinary in the dictionary. I couldn't believe it. There was a whole school on this. I stole it from the college resource center. I took it home. I told my parents that this is what I want.*

**Q / After you graduated, did you think right away that you were going to be a Chef?**

*A: I thought I was going to be a great saucier; that's what I wanted to do. I fell in love with that idea. During my externship, I worked at the sauce and soup station at the Hilton on Michigan Avenue in Chicago. All we did every day was support all the other venues by doing the soups and the sauces for every restaurant in this giant—I mean the hotel did thousands of covers a day. So there were a lot of sauces to be made and soups. And so it made sense that you could actually pursue a career in that type of specialty. That's what I wanted to do. I knew I wouldn't be a Chef right away. This was in the days that you would never call yourself a Chef until you were really a seasoned professional.*

*One of the reasons that I fell in love with going to culinary school was because I could live anywhere in the world and support myself. It was also a very big part of the romance of this industry. I could go to a Club Med and work, and I could travel the globe and get a job here and stay for a year then move on. And that was a very big part of the romantic aspect of being in the food industry—I could live anywhere and support myself.*

**Q / So what work experience did you get after graduating?**

*A: I was looking to build skills. I worked in a Scandinavian restaurant in Atlanta. I worked in a Russian restaurant in Atlanta. And then I worked in a continental cuisine restaurant. Smaller restaurants offered me the best opportunity to build skills. When I went to work for the smaller restaurants, they really appreciated me. I got to do the specials of the day. I cooked the line at night. I was a very loyal employee. That's where I really learned to cook. They gave me the opportunity to do all the cooking—the side dishes, the sauces, the entrées, the appetizers, everything. I even got a little exposure to desserts.*

*So, after working in Atlanta, I moved to Los Angeles. Los Angeles is a very big city, a very difficult place to find your way. And I had some very crappy jobs. I hit rock bottom when I was running a cookie store at a mall. I got a phone call from the woman who was a Pastry Chef at a restaurant where I had worked. She remembered that I wanted to bake and she told me that her Sous Chef just left for the summer, but he would be back in the fall. She asked me, "Do you want to learn how to bake right now?" And I said, you bet. So I quit the cookie store in a heartbeat and went to work for her. It was a frustrating summer. I didn't pick it up very easily but, by the end of the summer, she said I was a great Baker and a great Sous Chef. That was a big deal for me.*

*After that, I went to a brand-new restaurant in Redondo Beach which was about to open. I sold myself as a Sous Chef making simple desserts. And they were American-style desserts because that's what I love to create. I made specialties like chocolate-strawberry shortcake, linzer cookies, chocolate truffle, and seven-layer chocolate decadence. It was in the early 1980s, when there was a big renaissance in desserts. And I couldn't make enough of them. Within six months, I had a pastry assistant, and we were doing wine dinners every month with the pastries and special anniversary parties with pastries and getting written up in every newspaper, magazine, and guide in Los Angeles. It was a real exciting time for me. And the owners of the restaurant kept saying they wanted to open a bakery, but they clearly didn't have the money to do it.*

*So Eliot and I saved $20,000 and decided to do something on our own. I didn't really want to do retail bakery and sell cookies and wedding cakes. I wanted to sell my desserts to restaurants and businesses, and that was the premise for the business. I thought I was the first one to ever think of it, not* *realizing there was such a thing as a wholesale dessert company. Then during my first few weeks out there I started to learn that there was competition, so I was terrified. I thought, I can't do this. We would never survive.*

## Q / Did you rent a little store to start your business?

**A:** *Actually, it was a restaurant. But first, I baked six cakes in the apartment that we had at the time, and I took a slice out to my favorite restaurateurs. I asked them if I were to make these things, would they buy them? In those days, everybody had a Pastry Chef. And every single restaurant I went to said, "Yes, I would buy them, and I want two of those and three of those." And I said, well, I don't really have anywhere to make them because I was adamant that I could never cook professionally out of my house. So one of the people I visited said they had a restaurant in a nearby town that's not doing very well and is only open for lunch. They let me go in there at 1 AM, just be out by 10 AM. So I agreed, paid the rent, and that's how I got started. I would bake through the night, deliver the cakes in the morning, buy ingredients, and work on sales to build the business.*

## Q / Did you leave your job at that point to start your own business?

**A:** *Yes, and actually before we determined what direction to take the business, we traveled to Europe. Neither one of us had been to Europe, and I thought, well, I'm doing American-style desserts, and all the customers of this restaurant were saying have you ever had a strudel in Germany, or have you ever had a petit four in France, have you ever tasted any of these great desserts. And I hadn't. So I felt I didn't know what direction to take the company in. So we said we need to go to Europe. We're not going to have a vacation for a long time, and we never had a honeymoon. So we took a four-month trip to Europe and ate our way across the 11 countries of Western Europe. We went to every chocolate shop in Brussels, every strudel shop in Germany, and we went to the Hotel Sacher in Vienna. We just tasted all the regional pastries in every country. It was a magnificent trip.*

*I enjoy the American-style desserts with all the French and German techniques. I did make strudel, but it was my own style. We started the business with a whole renewed sense of pride in our own skills and style.*

## Q / Tell me about how you moved from supplying local restaurants with desserts to going national?

**A:** *A very big opportunity that changed our business was with a foodservice account where we walked in at just the right moment when a competitor had presented them with a chocolate dessert but couldn't come through with the pricing and the packaging. I was sitting in the lobby being stood up by the same Chef for the fourth time. And I'm sitting there with my box of cake—Pastry Chefs never travel empty-handed. So I had a cake in my hand, and it was a type of chocolate cake made like a soufflé with nowhere to fall. His assistant comes to tell me that he wasn't going to see me when she noticed the box and asked what was in it. I showed her my chocolate cake, and she asked me to sit right down. She went away and came back with a little tiny chocolate truffle cake and asked if I could make something like this. So I looked at it, and I could tell what it* *was by looking at it. I tasted it too. Because I had been baking cakes how many thousands of hours, I told her I could make the cake for her. I have a really good taste memory. I can taste something and figure out how to make it. She wanted the cake by Friday, which I did. Actually I think mine was better than hers. We agreed on details such as price and packaging design. We went national with that product in six weeks. We shipped pallet loads of product for the very first time. With Trader Joe's, we were doing fresh daily deliveries unwrapped—shrink-wrapped little cakes with labels on top. For foodservice, this had to be tray-packed, case-packed, marked, palletized, wrapped, and listed up, then brought to the distributor dock.*

*That changed the business. It was a struggle because I never knew what it meant to make a profit or keep books or hire employees or manage production schedules or any of that. I didn't know any of it. Eliot and I didn't know that part of the business. But we both had our own skills which sort of sifted out. We made a lot of mistakes. I did finally take classes at a local college on how to start a business. Taking this one small course at a community college taught me some very important basics that kept me in business—by taking liability insurance, by getting workers' comp, etc., by doing the right thing from day one.*

**Q / How many employees did you have at that time?**

**A:** *At that time, we probably had 35 employees. And that was a big deal. I was amazed at how fast we had grown.*

**Q / What skills are you looking for in a Chef that comes to work for you?**

**A:** *For Chefs to work in R&D, they need to have good knife skills, understand how to cook, have experience working a line and working with customers. I'm looking for people who actually have the foodservice experience, because how else can they help me create products for a retailer or a restaurant chain if they really don't understand the customers, the people who eat, how they eat, what they're eating, as well as how food is made? We bring in R&D Chefs who are great with the customers. They need to be able to do presentations demonstrating their culinary skills as well as to use PowerPoint and have photographs of menu items to give to customers. R&D Chefs are actually salespeople—they need to sell the products that they develop. They can't just stand there and cook the whole day and hope that maybe something hits.*

*The challenge of coming from a culinary background into our environment is also that you need to be very organized and detail-oriented. Those skills in a Chef can be hard to find. What makes it exciting for a Chef to work here is the opportunity to create new products for brand-name customers.*

**Q / What paths could a culinary student take to get into the R&D field?**

**A:** *I think there are a lot of different paths that one could go into to get into R&D. If a culinary graduate just wanted to go out and get a food science degree and do R&D, that path is open because someone with a culinary degree and a food science/technology degree would have no problem finding a job. That being* *said, in my opinion, I would recommend spending time learning the foodservice industry, because understanding how people eat, where they eat, what they're going to be eating, is a great way to connect to that customer. And you're going to need those skills in R&D and manufacturing. Additionally, running an organization where you have profit and loss responsibility, and you have an opportunity to manage people, all of that directly relates to a manufacturing facility, because you will need to control costs and supervise people.*

*I have two senior-level scientists and technologists who can't solve problems any better than our culinary creative team, because the Chefs' approach to things is really grass-roots and practical. R&D people are always focused on how to make that product hold that color, how to make it stay that thick, how to make it have that flavor. The Chefs and food technologists work great together. We're looking for the pure culinarian to work hand in hand with the pure technologist and personally we get a better result.*

### Q / What have you loved the most about growing your own business?

*A: What I have loved the most in the past is building the organization, finding the right people for the right job. That has always given me my greatest pleasure, as well as being able to be out networking with the industry and understanding the trends.*

### Q / And maintaining the quality that you started by yourself?

*A: Right. That hasn't given me pleasure though; that's been brutal. Now that the organization is built, I don't need to be involved. I have excellent people who can find their own people. The Chef can find more Chefs, and the CFO can manage his own finance department. So I'm not needed to do that recruiting anymore. Now I'm needed to do the new business development, thinking carefully about our strategies for the future. That is more my focus now.*

### DANIEL V. W. COUDREAUT, Executive Chef/Director of Culinary Innovation, McDonald's U.S.

### Q / What types of restaurants does Metro Media Restaurants run?

*A: Metro Media Restaurant Group has a number of chain restaurants: Bennigan's, Steak and Ale, Ponderosa, and Bonanza.*

### Q / Who do you report to?

*A: I report to the Senior Vice President of Marketing.*

### Q / What is a typical day of work?

*A: Essentially, I will come in and I'll hit my e-mails and voicemails. We have lots of market directors, franchisees, who will be communicating with me on a daily basis if needed. I act as a resource to these folks. Then, a large part of my day is spent in meetings trying to find out what the strategies are as far as where they are going with the direction of food.*

*Once I'm given that strategy in these meetings, then I will go out and start putting together the menu items and concepts that they want to have presented. They'll say, for example, that we really need to focus on chicken thigh meat because we have a contract for it. They will need x number of recipes to use chicken thigh meat in a food bar. So at that point, I develop specific recipes for that product. I spend a lot of time finding out what is the strategy, trying to coordinate all the different disciplines, whether it's purchasing, marketing, or operations. Then I go into the kitchen, develop the food, come out, take pictures, write up the recipes, work on the training materials with our training department, and move forward on a promotion. About 70% of my time is spent in our test kitchen, and a good amount of time is spent writing the recipes, costing out the recipes, and then there will be some time where I'm out actually in the field doing hands-on training shoulder-to-shoulder with our Market Directors, or what we call training the trainer.*

*I serve many masters, so to speak. I have to navigate the waters because Marketing wants the next wow product, Purchasing wants it to be cost-effective, and then Operations wants it to be very, very simple. So I have to find a product that nails each one.*

**Q/ How many hours do you work a week?**

*A: I would say I work anywhere from 45 to 50. I determine most of my schedule. I don't punch a clock or anything like that. I work as long as it takes me to get the job done. Sometimes I'm traveling on the weekends. In order to really understand what the restaurant is going to be on a Saturday night, you have to be there on Saturday night.*

**Q/ What was your career path to your current position?**

*A: I started as a dishwasher when I was about 14 years old. I worked in that restaurant as a breakfast and dinner cook, but mostly as a lunch cook. From there, I went into some catering at a conference center, and then worked as a line cook, sauté cook, grill cook, and shift leader at age 18 in a casual dining restaurant. Then I worked as a bartender and server. I was still doing catering on the side as a cook, so I was getting my feet wet in pretty much all of the aspects of restaurant management and operations. Then I moved to Manhattan and worked as a bartender/server. I moved over to a French bistro where I was the PM sous chef.*

*From that point, I enrolled in culinary school. I was going to get serious about this becoming my career, so I decided to go to school. For my externship, I opened up 64 Greenwich Avenue in Greenwich, Connecticut. When I graduated from school, I was recruited by Café Pacific, which is a highly rated restaurant in Highland Park Village, Dallas. I worked as their Executive Sous Chef, and then I went to the Four Seasons Hotel in Los Colinas as their Garde Manger Chef before being promoted to Chef de Cuisine of their fine dining restaurant. I was promoted again to Club Chef, and I oversaw four operations. I received most of my management experience when I was managing Sous Chefs and Cooks, anywhere from about 30 to 40 people.*

*I was on the fast track within the Four Seasons family, but I stepped back a bit and wondered where my quality of life was, my family, and where I was going with this. I investigated the opportunity to work in Research and Development, my current position. Although I felt I had no experience in this area, it allowed me to have a better quality of life, with holidays and weekends off. That was more important, although the R&D profession has proven to be very exciting and rewarding. Essentially, the sky's the limit with where I am now. There is a lot of opportunity to improve restaurant products, and that is why they need culinary professionals that have education and experience in all culinary fields.*

**Q / What do you like most about your job?**

*A: I like the people. Restaurant people are great people. They are hardworking and dedicated. I like being an educational resource. I like being able to supply answers that are based on a foundation of knowledge. And if I don't have the answer, I know where it can be found. So it helps them to make their job easier.*

**Q / What do you like least about your job?**

*A: The pace of decision making. Coming from restaurant operations, if something doesn't work today, you don't continue to put it on the menu. You basically change it immediately. When you are in a chain restaurant operation, it takes longer for those decisions to happen. I don't like indecision or bureaucracy. It's just the nature of what it is. I think that is probably what tends to frustrate me more. On a daily basis I'd like things to move quicker.*

**Q / Give some advice to new graduates who want to go into a career of your nature.**

*A: Don't be afraid to work. Have some patience. If you want to work in research and development, you need a diverse knowledge and experience base. You have to be able to draw upon a lot to get through a day. We are running restaurants. If you think chain restaurants are any different than independent restaurants, there are very subtle differences, but we are still putting food on a plate for a guest to eat. And it happens 365 days a year, so you have got to be able to have a base of knowledge that allows you to do that. I think hard work and different experiences will help anybody trying to get into this side of the industry. I don't think you should be closed to any experience. When I first went into the hotels, did I want to go into Garde Manger? Not really. But it was probably the best experience I ever had because it was the busiest kitchen in the restaurant. I learned how to hustle, organize my time, how to read banquet event orders. I learned all of that as well as being able to move up the ladder. So you have got to be open to getting your hands dirty and taking on the hard jobs sometimes.*

## ADVICE FROM A RECENT GRADUATE

Name:       Alexis Kaufmann
Education:    AOS, Culinary Arts, Culinary Institute of America, 2005

> *Learn from practical experience – take advantage of working with people from all different backgrounds, educations, and positions.*

# EXERCISES

1. Visit the website of the Research Chefs Association (www.culinology.org). What is the mission of the RCA? Who belongs to it? What are five benefits of being a member? How can you become certified as a Certified Research Chef or a Certified Culinary Scientist?

2. Compare and contrast the job duties of an Executive Chef in a restaurant to those of a Research Chef?

3. What other skills does an R&D Chef need, in addition to basic cooking skills? (Be sure to read the interview to answer this question.)

4. Visit the website of Lori Daniel's Two Chefs on a Roll (www.twochefsonaroll. com). Check out all the pages on the website. Would you like to work for a company such as this? Why or why not?

# CHAPTER 11

# Culinary Careers in Education

## Introduction

**TRAINING EMPLOYEES IS A PART OF EVERY CHEF'S JOB.** Chefs who find they really enjoy teaching can go on to become educators, either part time or full time. Chef educators work in many exciting work settings.

- Vocational-technical high schools—Most vocational-technical high schools offer training in the culinary arts. These high schools may be part of a regular high school or in a separate location. Nearly two-thirds of all high school graduates from vocational high schools enter some form of postsecondary program.
- Professional culinary schools—Many of these postsecondary institutions offer associate degrees or certificate programs in the culinary arts.
- Colleges and universities—These include both two-year colleges that award associate degrees and four-year institutions that award bachelors degrees.
- Cooking schools—Schools offer cooking classes mostly to adults who are not Chefs but simply want to improve their culinary knowledge and skills for use at home.

Most teaching careers (except within cooking schools) involve teaching students who are working or plan to work in the culinary field.

In all of these settings, educators often perform these job duties:

- Assess the learners' educational needs.
- Design and prepare classes.

- Teach classes, including food labs, using a variety of teaching methods and media.
- Demonstrate food preparation, cooking, and presentation techniques to groups of learners.
- Establish and maintain an effective learning environment.
- Respond to individual, cultural, and age differences in students.
- Work with learners individually and in groups.
- Motivate learners and inspire their trust and confidence.
- Advise learners in choosing courses, colleges, or jobs.
- Evaluate learners by preparing and grading tests and other assessment tools.
- Work cooperatively and communicate effectively with other learners, support staff, parents, and members of the community.
- Keep up to date with trends and developments in the culinary field.

This chapter introduces the life of a Chef educator.

# The Feel

## THE FIRST MONTH

As Chefs in a professional kitchen, we are predominantly teachers, mentors, and daily role models for our staff. After some time, many of our staff members move on to expand their careers and experience. Nothing is more rewarding than the phone calls you receive from past employees bringing you up to date on their lives and accomplishments. They feel like extensions of your family, as many of them worked for you in their early years, when they were still kids. The conversation usually ends with, "Hey, Chef, thanks for everything you taught me," or "Chef, you were right. I needed those basics to be successful, and now at work I blow everyone else away." Words like that are like music to our ears, for we are all teachers at heart. Our success is based on the success of those around us, trying to emulate what, when, and how we do our job. When the opportunity came my way to teach, I thought, "What a fine way to transfer my years of experience to the next generation."

The first month of being a Chef instructor in a professional culinary school was one of the toughest I have ever worked. Somehow or other, teaching didn't look so hard when I was a student. Teaching cooking and techniques in the food lab isn't nearly as hard as classroom teaching, yet I am constantly praying that everyone leaves the lab with all ten fingers intact and that my boss doesn't show up and see how disorganized and messy the kitchen has become. It's mind-boggling that some of my students have never cooked anything except microwave popcorn. But despite their primitive cooking skills, many of them seem to have a passion for foods and cooking, and that will help them (and me) along this path called education.

Teaching in the classroom is rough. I spend hours preparing my classes and putting together good visuals, yet I'm not entertaining enough to stop some students from instant messaging each other, playing computer games, or watching DVDs on their laptops. I'm supposed to be using new technology in the classroom to improve teaching, not developing ways to counter students' uses of technology that disrupt my class!

I know deep down that if I can run a professional kitchen, I can get a class of 19 students under control. I like the students; they are all good kids who are just looking for some direction. They keep me young in my thinking and constantly challenge my knowledge with their never-ending questions. If I don't have the answer, we either research it together or I have the answer for them the next day. I prefer to show them how to find answers than to tell them everything.

## MIDDLE OF THE SCHOOL YEAR

I survived the first semester, and thank goodness for the long break between semesters. The break allowed me to recharge my batteries and reflect on what worked and what didn't work, as well as to network with other Chef educators about how to deal with classroom disruptions and students who really don't want to be there. The school has rules, and there is a right and wrong way to properly handle a student not participating in a class. I want to make sure from the get-go that I do things correctly. All in all, the majority of students, now that I have gotten to know more of them, are good students who work and go to school. They may not get A's, but they are committed to cooking and at least have a sense of humor.

I'm not an academic student myself in the traditional sense. In fact, I chose this profession so I could be more hands-on. The truth is, there is a portion of our profession that is academic and can expand with the areas we choose. For the most part, as your commitment and dedication grow for your profession, your academic skills develop from the passion you possess for cooking. If we try to incorporate hands-on activities in the midst of the classroom, fewer students drift off. One of the reasons we went into this work is because it is fast-paced, always changing, interactive—which is what people like us love. We definitely aren't loners. We enjoy stimulating conversations, interfacing with other people, and, most important, we need to be busy. Luckily, I've discovered that the problems I have in the classroom occur in every corner of this institution, so I've gotten lots of ideas from the other instructors here. They are a great resource, especially if you make them lunch or dinner.

## END OF THE SCHOOL YEAR

Somehow all the blood, sweat, and tears I put into my first year of teaching disappeared as I watched my students graduate. I felt a sense of accomplishment I have never felt before. I know I have made a positive impact on the lives of many of these young culinarians, and I have a great feeling of pride whether they thank me or not.

There is a significant parallel between teaching and running a professional kitchen. You must use the same patience and understanding when instructing a student as you would with a new employee. The realities of a lab are not much different from a typical day of kitchen instruction. The pressures may be different, but not the people. The students as well as the employee must have a sense of your expectations of them. They need to understand that the Chef-educator is there to guide and inform them as they learn and progress, but responsibility for the result is ultimately their own. Lecturing is another story altogether. You must recognize the different learning styles of the students. Some are visual learners; others are hands-on learners or verbal learners.

A teacher is like a conductor of an orchestra; he or she must be gentle and instructional to the violins, then turn with command and urge on the drums while confidently creating a flow and balance for the trumpets. This is the most challenging aspect of the job, while at the same time it is the most rewarding. I now see that teachers can have a long-lasting influence on their students. We can affect how students learn, what they learn, the quantity and quality of what they learn, and the ways they will interact in the future.

Having so much power can be a little scary, but I believe my understanding, passion, and commitment in being a Chef has had a positive influence on these students. I'll be coming back to teach next year. Yes, it will still be a challenge, but I also consider it a privilege.

## A Day in the Life

Arriving for work by 7:30 AM every day at the community college, the first thing I do is check my email, and then by 8 AM I check with the lab assistant to make sure I have all the foods needed for my 1 PM lab class. Everything is in, so I finish preparation for both the lab and my 10 AM lecture on baking yeast breads. By 9 AM I have a few students in the office. Some have questions; some just want to say hi and chat for a few minutes. I love the interaction with the students. At 9:30 AM I have an appointment with a student in another major who wants to switch to culinary arts. I review his transcript, tell him what additional courses he will need, and give him an idea of what cooking and being a Chef is like.

Now it's on to my 50-minute class on bread baking. Teaching lecture classes these days is a challenge because you can't lecture in the traditional sense without losing most of the class. To be successful, I have to come up with great visuals for my presentations, find ways to get students to interact with the material and each other, and shift gears when I see I've lost their attention. Luckily, I spend many more hours in the kitchen for food labs doing hands-on activities. Most students enjoy the hands-on work best. In any case, they get to apply what we talked about in the classroom setting.

After answering student questions at the end of class, I dash to an 11 AM department meeting. We meet once a month to discuss college issues, but we also update each other on what each program is doing. As in some other community colleges, the culinary arts program is in the business department, so in addition to interacting with the one other full-time Chef instructor, I work with people who teach legal assistant, administrative assistant, and other business classes. My boss, the Dean of the Business Studies Division, runs the meeting and shows us how to fill out the forms requesting new equipment for our programs. As always, money is tight, and you have no guarantee you will get what you need.

The meeting goes until 12:20 PM, at which time I grab something to eat in the student center, change into my Chef's outfit, and get into the kitchen lab. The lab assistant has laid out all the supplies we need for the day, which includes both ingredients to make bread dough from scratch and prepared bread dough. To start the lab, I demonstrate how to make the yeast bread dough and then supervise the students in making their own dough. While the dough rises, I use bread dough that has already been fermented to demonstrate scaling, rounding, benching, makeup, and panning. The students keep busy the rest of the class by using the prepared bread dough to make loaves, and then use the dough they prepared themselves to make a variety of hard rolls. Toward the end, they work in teams to make sure the lab is clean and everything is put away. At 5:05 PM, I release the students to go home. They are proud of their new bread-making skills.

Once I finish responding to the emails and phone messages, it's time to check what's on my calendar for tomorrow. Luckily I have no college meetings, so I should be able to finish grading a test I gave yesterday. But the high point of the day will be when I watch the Quantity Foods class make and serve the foods at a college event. I can't wait!

# The Reality

In any setting, seeing learners gain knowledge and develop new skills is highly rewarding. However, teaching has its share of frustrating moments, whether working in cooking facilities that need to be updated or dealing with unmotivated or disruptive students. Students, and teachers too, come from varied backgrounds. With growing minority populations in most parts of the country, it is important for teachers to work effectively with a diverse student population.

This section discusses the reality of teaching in cooking schools, vocational high schools, and postsecondary schools.

## COOKING SCHOOLS

Working as a cooking instructor in a cooking school can be a lot of fun. Learners pay to be there, so they are motivated to learn and are invested in getting something out of your class. Most cooking school jobs are filled by freelancers who are paid by the class. Cooking school instructors may specialize in certain types of classes, such as baking or cooking for children. The highlight of cooking classes is the recipes, so recipes must be carefully chosen, concisely written, and absolutely correct in measurements, directions, temperatures, and timing. The instructor or the school may be responsible for ingredient shopping. Before class starts, instructors must have the lab, including food and equipment, ready for the learners. During class, instructors have a lot to do: explaining and demonstrating foods and cooking, supervising food preparation of different dishes by various groups, answering learners' questions, having tastings, setting up meals.

## VOCATIONAL HIGH SCHOOLS

To teach in a vocational high school, you must be licensed by the state in which the school is located. In many states, vocational teachers (also called career technology teachers) must meet many of the same requirements for teaching as other secondary school teachers. However, because knowledge and experience in a particular field are important criteria for the job, some states license vocational education teachers without a bachelor's degree, provided they can demonstrate expertise in their field. A minimum number of hours in education courses may also be required. For example, in Pennsylvania you can get an intern certificate by enrolling in an approved college program and meeting program requirements. To receive your permanent certification, you must complete three years of satisfactory teaching in a vocational high school as well as 60 college credits.

Most states have tenure laws that prevent teachers from being fired without just cause and due process. Secondary teachers may obtain tenure after they have satisfactorily completed a probationary period of teaching, normally three years. Tenure does not guarantee a job, but it does provide some security.

Including school duties performed outside the classroom, most teachers work more than 40 hours a week. Most vocational teachers work traditional ten-month school years, with a two-month summer vacation. Teachers in districts with a year-round schedule typically work eight weeks, are on vacation for one week, and have a five-week midwinter break.

Area employers frequently provide input into the curriculum and offer internships to students. Culinary arts teachers must be able to play an active role in building and overseeing these partnerships. For example, teachers may work with the National Restaurant Association Educational Foundation to use its ProStart® program. This state-based program for high school juniors and seniors links two years of classroom curriculum with a mentored worksite experience. Over 40 states currently have schools that use the ProStart® program.

## POSTSECONDARY SCHOOLS

Postsecondary schools include professional culinary schools, colleges, and universities. In the past, students entered postsecondary schools right after high school and attended full time. Nowadays, postsecondary faculty work with an increasingly varied student population made up of growing shares of part-time, older, career change, and culturally and racially diverse students.

Postsecondary instructors prepare classes, grade papers, attend departmental and faculty meetings, and keep abreast of trends and developments in the culinary field. The teaching load is usually heavier in two-year and culinary schools than in four-year institutions. Faculty members are often involved in setting up and supervising internship programs for students, and they provide students with information about prospective employers. Most faculty members also serve on one or more academic or administrative committees that deal with the policies of their institution, academic issues, curriculum, budgets, equipment purchases, and hiring.

Postsecondary teachers usually have somewhat flexible schedules. They must be present for classes and for meetings. Most establish regular office hours, usually three to six hours per week. Otherwise, they work on course preparation, grading, study, research, and other activities.

Some postsecondary teachers teach night and/or weekend classes. This is especially true for teachers at schools with large enrollments of older students who have full-time jobs.

Depending on the school and the position, full-time contracts may require you to work 9 months each year or 12 months each year. About three out of ten college and university faculty worked part time in 2002. Part-timers, known as adjunct faculty, often teach in addition to doing their regular job. Many come into the college or university one evening a week during the semester to teach one course. Part-timers generally get no benefits; their pay varies from $1,000 to over $3,000 or more for a three-credit course.

Training requirements for culinary schools vary, but in general, teachers need at least an associate degree plus work experience. More schools are recommending a bachelor's degree or higher, and this may become a standard requirement. Certification by the American Culinary Federation may also be required.

In two-year colleges, master's degrees are preferred and often required, although some jobs, especially part-time jobs, may accept a bachelor's degree. Many two-year colleges increasingly prefer job applicants with some teaching experience.

Four-year colleges and universities usually consider doctoral degree holders for full-time, tenure-track positions, but they may hire master's degree holders or doctoral candidates. Most college and university faculty are in four academic ranks (from highest to lowest): professor, associate professor, assistant professor, and instructor. Most full-time faculty members are hired as instructors or assistant professors. A smaller number of additional faculty members, called lecturers, are usually employed on contract for a single academic term and are not on a tenure track.

For faculty, a major step in the traditional academic career path is attaining tenure. New tenure-track faculty are usually hired as instructors or assistant professors, and must serve a period—usually seven years—under term contracts. At the end of the period, their record of teaching, research and writing, service, and overall contribution to the institution is reviewed. Tenure is granted if the review is favorable. Those denied tenure usually must leave the institution. Tenured professors cannot be fired without just cause and due process. Tenure protects the faculty's academic freedom—the ability to teach and express opinions without fear of being fired for advocating unpopular ideas. It also provides financial security for faculty. Some institutions have adopted posttenure review policies to encourage ongoing evaluation of tenured faculty.

The number of tenure-track positions is expected to decline as institutions seek flexibility in dealing with financial matters and changing student interests. Institutions will rely more heavily on limited-term contracts and adjunct faculty. In a trend that is expected to continue, some institutions now offer limited-term contracts to prospective faculty—typically two-, three-, or five-year full-time contracts. These contracts may be terminated or extended when they expire. Institutions are not obligated to grant tenure to the contract holders. In addition, some institutions have limited the percentage of faculty who can be tenured.

For most postsecondary teachers, advancement involves a move into administrative and managerial positions, such as departmental chairperson or dean. At most four-year institutions, a doctoral degree is necessary to advance.

# Professional Organizations

Please see Appendix A for more information on each of the following professional organizations for culinary and foodservice professionals working in education.

American Culinary Federation

Association for Career and Technical Education

Foodservice Educators Network International

International Association of Culinary Professionals

International Council on Hotel, Restaurant, and Institutional Education

# Interview

**THOMAS RECINELLA, CEC, AAC, Assistant Professor, State University of New York at Delhi**

**Q / How many students are in the Hospitality Management Program?**

*A: There are over 300 students in hospitality management. Within hospitality management, there are majors in travel and tourism, culinary arts, hotels and resorts, and restaurant/foodservice management. About 140 are in culinary arts, and they are split between those going two years for an Associate in*

*Applied Science (AAS) and those going four years for a Bachelor in Business Administration (BBA). All majors offer AAS and BBA degrees. We offer the only BBA in culinary arts in the state of New York.*

**Q / How is the department you teach in organized?**

*A: We have a division Dean who splits her time between the hospitality department and the business department. All faculty report directly to her. We have two secretaries, and our Dean reports to the Vice President of Academic Affairs.*

**Q / Besides preparing and teaching classes, what other responsibilities do you have?**

*A: I do a lot more besides teach classes. I:*

- Coordinate all culinary student activities, culinary competitions, fundraisers, and any dinners that we put on for the department.
- Advise the culinary club, which is a philanthropist organization called the Delhi Escoffier Club.
- Coach the culinary teams and chair all culinary competitions/contests held on campus.
- Order all of the food for culinary arts, including developing specifications and negotiating prices.
- Supervise the making and harvesting of the ice blocks from the Clinebell machine (for making large ice blocks) to be used for ice carving.
- Advise 45 students in their academic careers.
- Handle job placement for summer internships.
- Design and spec new lab spaces and equipment as needed.
- Work on search committees when needed.
- Serve on other committees when needed.
- Serve as the division's representative on the college senate.
- Write new courses and revise old curriculum as necessary.

*To date, I have written five new courses which are used in our program. I stay pretty active in virtually every department endeavor.*

**Q / What is it like to be a teacher in the college classroom these days?**

*A: It is challenging, especially for Chef educators. With career changers and traditional students, we need to be able to relate to both. All students learn differently, so it is very challenging, to say the least, when speaking of a lecture. Lab is a little different. It is easier for me anyway to guide a struggling student through lab than it is in the classroom setting. The lab is really no different from a normal kitchen situation. The patience and the willingness of the Chef to work with the employee is the same in a lab situation. The student must know that there are expectations and what they are. Then they must know and feel that the Chef will help to guide them to those expectations, but the responsibility is ultimately theirs. In lecture, it is quite different. Some students are visual learners, while others learn by doing. There seem to be more students with special needs these days than there were when I went to college, or maybe we are just doing a better job of identifying and helping them than we used to. Learning disorders and attention disorders are very prevalent. Diversity of course is a big issue. A teacher in today's classroom is a lot like a maestro. The key is to treat everyone with respect and consideration and hope that your example is transferred to the entire class.*

**Q / Describe a typical workday.**

*A: In between 7:00 and 7:30 AM, I prep for class. I also compile orders for the restaurant class. Each day is different depending on the semester load. I teach 26 contact hours currently, meaning that I am in class or lab 26 hours each week. Labs are four hours long. My club meets once a week. I leave school by 7:30 PM, except on Fridays when the restaurant is open, I am not done until 11:00 PM. I guess my days are not typical. But that is what I like. I am often in on Saturday, especially when our team is practicing.*

**Q / What do you like most about your job?**

*A: I like the students. They keep me thinking young, and they challenge me not to grow stale and uninformed. I like being at a small school where I can do it my way. And my way has worked. (I like that best.) I like knowing my students' names, all of them. I like knowing that Mike comes from Binghamton and has a dog named Buddy and a sister in the army. I like the feeling when I leave at night or watch one of my students graduate that they are better prepared than I was when I left school, and they got there with my help. I like making a difference. I love to work, and the students give me a lot of work to do.*

**Q / What are the challenges in your job?**

*A: I am not an academic person in the traditional sense. I am a Chef and a cook. I like to make things happen. In academia, sometimes things move way too slow. I am a 6 AM to 10 PM guy working in a predominantly 9-to-5 world. That can be frustrating at times. A recent challenge has arisen with our continued success. We are constantly asked to take more students. I refuse to do it. Our strength is our instructor-to-student ratio. I will not compromise the integrity of that ratio. My boss supports me in this as well. The pressure comes from above her. But we deal with it. The only way we can take any more is to build a new facility. I sometimes get frustrated with those professors that I encounter that take the students for granted. They forget why they are here and why they have a wonderful job.*

**Q / What led you to teaching? What was your career path?**

*A: I started at 14 as a dishwasher and line cook, working with my sister in a greasy spoon in Livonia, Michigan. I worked there and a similar place for six years and then went to college for culinary arts. I have been a Technical Chef in a large university dining service and a Chef de Cuisine for a private club as well as a 2,200-seat banquet facility. I worked in large-volume à la carte restaurants doing 900 to 1000 covers.*

*I am really a line cook at heart. My son's birth led me into teaching part time in Michigan to make some extra money, and I took a liking to it. I pursued it full time here at Delhi by taking over the struggling culinary program five and a half years ago. The program was in jeopardy, and I was their last resort.*

*I also owned and operated several of my own businesses. Most recently, my wife and I had our own restaurant for three years here in New York. I have much experience in off-premises catering for 100 to over 5,000 guests. I also worked in several smaller family Italian restaurants. I now run a very private and exclusive club in the Catskills in addition to my teaching. I do all of the cooking and baking, and my wife works with me. It is very important to stay in tune with cooking and running a kitchen. I think it is a disaster when Chef educators leave the kitchen entirely. They cannot possibly be effective teachers without engaging in some kind of cooking outside of school. Labs are fine, but they just do not cut the mustard for keeping an instructor up on their own cooking ability.*

**Q/ What professional associations for teachers do you belong to?**

*A: I belong to the American Culinary Federation and the National Ice Carving Association. I follow Council on Hotel, Restaurant, and Institutional Education, and I read a great deal.*

## ADVICE FROM A RECENT GRADUATE

Name:        John D'Arcangelo
Education:   BPS, Culinary Arts Management, Culinary Institute of America, 2003

> *Establish at least one mentor, someone who will support you but will also push you, even when you have had enough.*

## EXERCISES

1. Would you consider working in education either full time or part time? Why or why not? If so, which type of school or college would you like to work in and where?

2. Visit the website of a cooking school such as Tante Marie's Cooking School (www.tantemarie.com) and look at the classes offered. Do they offer day, evening, or weekend classes? What types of classes seem popular — basic skills classes, international cooking?

3. If you worked full time in a postsecondary school, what benefits would professional organizations such as ACF, I-CHRIE, or FENI offer you?

4. Explain why you agree or disagree with Tom Recinella's following statement: "I think it is a disaster when Chef educators leave the kitchen entirely. They cannot possibly be effective teachers without engaging in some kind of cooking outside of school. Labs are fine, but they just do not cut the mustard for keeping an instructor up on their own cooking ability."

5. Find out from one of your instructors why he or she became an educator and what he or she likes and dislikes about the field.

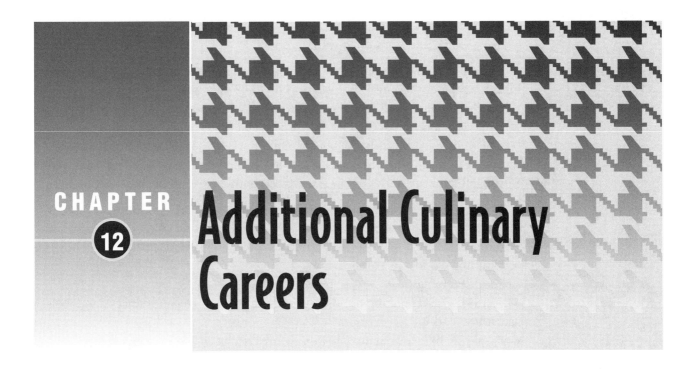

# Additional Culinary Careers

## Introduction

**THERE ARE MANY INTERESTING CAREERS FOR CULINARIANS IN ADDITION TO THOSE DISCUSSED IN THE LAST 11 CHAPTERS.** This chapter discusses what it is like to work in the following fields.

◆ Personal and private Chefs
◆ Food stylist and food photographer
◆ Food communications
◆ Public relations
◆ Food purchasing managers
◆ Celebrity Chef

Without a doubt, we didn't cover all the jobs that a Chef could venture, but we have tried to show you the venues where the large majority of Chefs work.

## Personal and Private Chefs

Personal Chefs run their own businesses to provide customized home-cooked meals for busy individuals and families. They help individuals and families who:

◆ Don't have enough time to plan and cook meals.

◆ Want to eat healthier meals.

◆ Are giving a party but don't want to cook.

The number of personal Chefs has been growing. Some individuals are going into this segment because they want to be their own boss, or they want to make a comfortable living while working flexible daytime hours.

What does a personal Chef do? Most personal Chefs own and operate their own business and have a number of clients. Once a week, they usually prepare Monday through Friday meals (often just entrees and side dishes) for one client each week. During a typical week, he or she is involved with any or all of these activities:

◆ Conducting an assessment with potential clients to learn about what foods they like, whether or not they have any allergies or special dietary needs, and so on.

◆ Planning weekly and other menus.

◆ Food shopping.

◆ Setting up and cooking for clients, including preparing containers and labels with reheating instructions.

◆ Preparing bills for clients and depositing checks.

Although it is possible to start your business with clients you already know, you will at some time have to market your services to gain more clients.

Whereas a personal Chef has a number of different clients and is self-employed, a private Chef has only one client, and is usually an employee of the client. Private Chefs prepare three meals per day and often live with the client, as well as travel with the client.

To learn more about being a personal Chef, read the following interview with Candy Wallace, Executive Director of the American Personal Chef Association (APCA).

# Interview

**CANDY WALLACE, Owner/Operator, The Serving Spoon**

**Q/ In addition to being the Executive Director of APCA, do you run your own personal Chef business?**

*A: Yes, I do. I am the owner and operator of The Serving Spoon, a personal Chef service in San Diego, California.*

**Q/ What types of job duties do personal Chefs perform each day, each week?**

*A: We counsel clients and design menus specifically to meet their dining and nutritional needs. If a personal Chef is amply qualified and trained, he or she may discuss medically specific diets. Generally speaking, meals are prepared in the clients' homes. This is helpful from a sanitation and transportation point of view because the food is prepared and immediately stored at the proper temperature.*

**Q / How many hours do you work each week in your business?**

*A: Like many other personal Chefs, I usually work with one client each day. I generally work five hours each day, about 25 hours weekly.*

**Q / What can a personal Chef expect to earn by cooking for one client each day, Monday through Friday?**

*A: About $60,000 to $75,000 annually, but there's a potential of $100,000.*

**Q / What was your career track to this position?**

*A: I worked in restaurants and have a lot of experience in catering businesses. I also went to college and got a degree in political science. I moved to San Diego, and owned a catering business called Taste, which did primarily dinner parties. In 1992, I realized that the catering grind was too much and I wanted to control more of what I was doing. I am a big entertainer, so I got all my friends together and told them, "I will support you and your families every day with fresh meals," and I did just that. I got a great deal of exposure and started to get calls from people who wanted my services as well as calls from people who wanted to start this type of business.*

**Q / What would you have done differently in your career path if you could do it again?**

*A: Since my only formal culinary training was continuing education courses, perhaps I would have gone to culinary school rather than university.*

**Q / Has any investment made in your career paid off?**

*A: What has helped me a lot is the fact that I have traveled a lot internationally and domestically.*

**Q/ What do you like most about your job?**

*A: I make an enormous contribution to the quality of life of my clients by providing great nutritious meals and removing stress from their lives. Personal Chefs bring families back to the dinner table.*

**Q/ What do you like least about your job?**

*A: Paperwork and accounting.*

**Q/ What advice do you have for new graduates in choosing a career like yours?**

*A: Go to school, learn the skills, and get some varied foodservice experience, because once you are in this business, you are on your own. This business should be entered once you are ready, individually and professionally, to operate in a self-directed environment.*

**Q / Is it true that personal Chefs can become certified through the American Culinary Federation (ACF)?**

*A: Yes, there are several certifications available: Personal Certified Chef (PCC) and Personal Certified Executive Chef (PCEC). Also, Certified Culiniarian is being added for personal Chefs. Requirements for these certifications are available on the ACF website, www.acfChefs.org.*

# Food Stylists and Food Photographers

Look at the front cover of any food magazine and you'll see the work of food stylists and food photographers. Food stylists and photographers work together to produce those beautiful, glossy food photos. But you need to be cautious about eating the food they photograph. Food stylists are like makeup artists: They use all types of tricks to make food look fresh and delicious, including some that make it inedible. For example, instead of using powdered sugar on top of a quick bread, a food stylist may use talcum powder because it won't melt. Another trick is to substitute motor oil for pancake syrup because it looks thicker.

Food stylists arrange food to be photographed professionally. This is a creative job that involves working with food photographers, photo editors, art directors, and others involved in print publications such as newspapers, magazines, and cookbooks. Food stylists also prepare food for advertisements, movies, television shows, and theater productions.

Here are tasks food stylists typically do:

- Meet with clients to discuss what is to be photographed and how, including props.
- Shop for food and equipment needed for the shoot or presentation.
- Cook or bake the food items.
- Arrange the food over the course of the shoot or presentation.
- Maintain a toolkit to use when styling foods, including dishes, glasses, and other items that can be used as props.

A stylist's toolkit also includes the tools needed for most of the tricks that keep food looking delicious and fresh—paintbrushes, skewers, tiny spatulas, and eyedroppers. Also, a food stylist may spray ArmorAll® on a tortilla so it doesn't dry out.

A culinary background, meaning both schooling and cooking experience, is crucial to succeed as a food stylist. In addition, food stylists need training in art and photography. An artistic background helps a stylist use colors, props, camera angles, and lighting to make food look great. Good communication skills help a stylist interact with clients, editors, producers, and other people he or she will work with to set up the stage. Because many food stylists work on a freelance basis, they also need some business background so they manage their business and finances well.

Besides having a background in foods and cooking, art, and business, a food stylist must be able to work well with clients and handle deadline pressure. Food stylists have to be flexible when the client wants to change something or has a strong opinion about a look. A food stylist may prepare a dish ten times before a good shot is taken, or freshen up a dish many more times, so patience is important. On photo shoot days, hours are usually quite long—ten hours or more.

To break into this field, most people start by getting a job as an assistant stylist. This is a good way to learn food styling techniques as well as photography. Most culinary schools offer classes in food styling; these are helpful, but they cannot replace hands-on experience with food stylists and photographers.

To start a career as a food stylist, you must do a lot of networking, because much of this business is by word of mouth, and you need to develop a strong portfolio of sample shots. Full-time food stylist positions are rare, so most stylists work as freelancers. Potential employers of both full-time and freelance stylists include advertising agencies, food companies, restaurant chains, food magazines, and television networks. Los Angeles and New York are two of the biggest places where food stylists are needed because that's where many advertising agencies, films, and television shows and commercials are produced.

Although the job outlook for food stylists is steady, there is a lot of competition for the jobs, in part because the work is fun and the hours are good. Full-time food stylists make from $30,000 to $50,000 or more each year. Freelancers make anywhere from $300 to $1,000 or more per day for a photo shoot, depending on the type of job, location, and their experience. Self-employment allows for greater autonomy, freedom of expression, and flexible scheduling. However, income can be uncertain and the ongoing and time-consuming search for new clients can be stressful.

Like food stylists, food photographers work most often as freelancers. Less than half their time is normally involved in photographing foods. Other tasks include researching new clients such as art directors and magazine editors, talking to clients about upcoming assignments, preparing for a shoot, estimating and negotiating fees, getting supplies, developing and delivering prints, bookkeeping, paying bills, return phone calls and emails, and so on.

The food photographer uses a variety of tools to create a mood or feel in a picture. He or she uses natural and commercial lighting to help paint a picture, almost like an artist does. For example, a photographer might use a wooden block to create a shadow for the feel of a candlelit meal.

To break into this exciting field, you might start work as an assistant to a photographer or even assist a food stylist. Like food stylists, most food photographers work as freelancers, and the field is very competitive. Freelance photographers can make upwards of $1,000 per day for a shoot, but keep in mind that they need a lot of money for equipment to start up their business. They also need to be well versed in digital photography and software programs like PhotoShop. Photography is a highly technical field, and attention to details is crucial.

Food stylists and food photographers often belong to the International Association of Culinary Professionals (IACP). IACP is a group of approximately 4,000 food professionals from over 35 countries. IACP provides continuing education, networking, and information exchange for its members. See Appendix A for more information about IACP. Food photographers may also be members of the Advertising Photographers of America or the American Society of Media Photographers.

# Food Communications

Culinary professionals can work in communications as food writers, food editors, and cookbook authors. Writers and authors develop original writing for books, magazines, trade journals, online publications, company newsletters, radio and television broadcasts, motion pictures, and advertisements. Journalists write articles mainly for newspapers and magazines. Editors select material for publication or broadcast. They review and revise a writer's work for publication or dissemination.

Major newspapers and newsmagazines usually employ several types of editors. The executive editor oversees assistant editors who are responsible for particular subjects, such as food and cooking. The food editor determines the subject or content of the food pages, which appear once a week in many newspapers or in each issue of a magazine. Depending on the publication, the food editor may use staff writers to write the articles or hire freelance food writers. Freelance food writers find work by writing letters to editors that outline potential article ideas; they also show samples of their published work to editors. If a freelancer's idea is approved by the editor, the writer writes and submits the article for publication.

Food articles in newspapers and magazines have become much more sophisticated recently. Whereas in the past it was routine to find articles on 25 ways to use up Thanksgiving turkey, nowadays you can read about how people all around the world grow their food, cook, and eat. Newspapers and magazines have discovered that readers long for stories and information on cooking and eating.

A good place to begin a career in food writing is to offer to write a food and cooking column for a local newspaper. Many communities have local weekly papers that are good training grounds for food writers. Once you have a number of clips—copies of your published articles—you can use them like a portfolio when you submit queries to magazines and bigger newspapers.

Becoming a cookbook author is a lot different from getting published in your local paper, and a lot more competitive. Further, publishers are usually looking for authors who are well known, experienced, and expert in their field. To get a book contract, cookbook authors write a book proposal detailing what the book will be like, including a table of contents and a sample chapter. Because the cookbook business is so competitive, most authors are represented by a literary agent. The agent, who usually keeps 15% of the eventual royalty earnings, tries to sell the book idea to publishers. Royalties are usually from 7 to 15% of the book's retail price.

To be a writer, editor, or journalist, you must be able to express ideas clearly and logically, and you should love to write. Creativity, curiosity, a broad range of culinary knowledge, self-motivation, and perseverance also are valuable. Most food writers and authors work as freelancers, so they need to know how to run their own business.

Employment of writers and editors is expected to grow about as fast as the average for all occupations through the year 2012. The outlook for most writing and editing jobs is expected to be competitive, because many people with writing or journalism training are attracted to the occupation.

Employment of salaried writers and editors for newspapers, magazines, and book publishers is expected to increase as demand for these publications grows. Magazines and other periodicals increasingly are developing market niches—including food and cooking. Businesses and organizations are developing newsletters and websites, and more companies are experimenting with publishing materials directly for the Internet. Online publications and services are growing in number and sophistication, spurring the demand for writers and editors, especially those with Web experience. Advertising and public relations agencies, which also are growing, should be another source of new jobs.

Many Chefs working in this field are paid on a freelance or royalty basis. Median annual earnings for salaried writers and authors were $42,790 in 2002. Many Chefs working in food communications are members of the International Association of Culinary Professionals (see Appendix for more information).

# Public Relations

Public relations managers work in hotels, restaurants and restaurant chains, cruise lines, supermarkets, universities, healthcare, and other businesses. The reasons why Chefs and restaurateurs hire public relations managers or agencies are as varied as the types of food they serve. Chefs and restaurateurs use public relations managers to help open a new restaurant, reposition an existing restaurant, raise both media and consumer awareness of the restaurant and/or Chef, or make customers aware of a new menu or service. Public relations managers use every available communication medium to place news and general interest stories about the company in local and broader publications with the aim of attracting customers. For example, public relations managers develop media relationships by keeping restaurant reviewers and food editors informed about special events.

Public relations managers are also involved in drafting press releases. Many radio and television special reports, newspaper stories, and magazine articles start in public relations as press releases. Public relations managers also work to uphold the image of the foodservice after a problem, such as a robbery or food-borne illness, strikes.

Public relations managers also evaluate advertising and promotion programs for compatibility with public relations efforts and serve as the eyes and ears of top management. They observe trends that might ultimately affect the foodservice and make recommendations to enhance the foodservice's image based on those trends.

Public relations managers may assist foodservice executives in drafting speeches, arranging interviews, and maintaining other forms of public contact; oversee company archives; and respond to information requests. In addition, some handle special events that introduce new products or other activities the foodservice supports in order to gain public attention through the press without advertising directly.

In public relations firms, a beginner might be hired as a research assistant or account coordinator and be promoted to account executive, senior account executive, account manager, and, eventually, vice president. A similar career path is followed in corporate public relations, although the titles may differ. Some experienced public relations specialists start their own consulting firms.

There are no defined standards for entry into a public relations career. A college degree combined with public relations experience, usually gained through an internship, is considered excellent preparation for public relations work. Some entry-level public relations specialists have a college major in public relations, journalism, advertising, or communication. However, many public relations specialists in foodservice have backgrounds and degrees in the culinary/foodservice field, along with an understanding of public relations and marketing, writing and editing skills, and excellent people skills. Creativity, initiative, good judgment, and the ability to express thoughts clearly and simply are essential. Decision making, problem-solving, and research skills are also important. People who choose public relations as a career need an outgoing personality, self-confidence, an understanding of human psychology, an enthusiasm for motivating people, and a passion for the culinary field. They should be competitive, yet able to function as part of a team and open to new ideas.

Some public relations specialists work a standard 35- to 40-hour week, but unpaid overtime is common. Occasionally, they must be at the job or on call around the clock, especially in an emergency or crisis. Public relations offices are busy places; work schedules can be irregular and frequently interrupted. Schedules often must be rearranged so workers can meet deadlines, deliver speeches, attend meetings and community activities, or travel.

Median annual earnings in 2002 were $60,640 for public relations managers. Salary levels vary substantially depending on the level of managerial responsibility, length of service, education, firm size, location, and industry. With frequent openings of new restaurants, as well as the rise of the celebrity Chef, demand for public relations professionals will remain steady.

The Public Relations Society of America accredits public relations specialists who have at least five years of experience in the field and have passed a comprehensive exam. They receive APR status—Accredited in Public Relations. The International Association of Business Communicators also has an accreditation program for professionals in the communication field, including public relations specialists. Those who meet all the requirements of the program earn the Accredited Business Communicator (ABC) designation. Candidates must have at least five years of experience in a communication field and pass a written and oral examination. They also must submit a portfolio of work samples demonstrating involvement in a range of communication projects and a thorough understanding of communication planning.

# Purchasing Managers

Purchasing managers seek to obtain the highest-quality food, beverages, and supplies at the lowest possible purchase cost for their employers. Chefs who especially enjoy the food purchasing part of their job may want to specialize in this area by becoming a purchasing manager for a large hotel or restaurant, hotel or restaurant chain, managed services company, cruise line, or other food-related business.

Purchasing managers determine purchase specifications, choose the suppliers of the product, negotiate the lowest price, and award contracts that ensure that quality, pricing, quantities, and product identification remain consistent. In order to accomplish these tasks successfully, purchasing managers study sales records and inventory levels of current stock, track seasonality and trends, identify foreign and domestic suppliers, and keep abreast of changes affecting both the supply of and demand for needed products and materials.

Purchasing managers evaluate suppliers on the basis of price, quality, service support, availability, reliability, and selection. To assist them in their search for the right suppliers, they review catalogs, industry and company publications, directories, trade journals, and trade shows. Much of this information is now available on the Internet. Purchasing managers research the reputation and history of suppliers and may advertise anticipated purchase actions in order to solicit bids. At meetings, trade shows, conferences, and suppliers' plants and distribution centers they examine products and services, assess production and distribution capabilities, and discuss other technical and business considerations that influence the purchasing decision. Once the purchasing manager has gathered all of the necessary information, he or she places orders and awards contracts with those suppliers who meet the purchaser's needs. Contracts often are for one year and usually stipulate the price or a narrow range of prices, allowing purchasers to reorder as necessary.

# Celebrity Chefs

Celebrity Chefs enjoy something most Chefs never have: a national audience. Take a look at Wolfgang Puck. He cooked for years without an audience until his business took off and grew to include 59 restaurants, six cookbooks, a television show, prepared foods, cookware, and a syndicated news column. Americans' appetite for celebrity Chefs, and especially their cooking shows, has grown tremendously. The Food Network has doubtlessly helped launch many Chefs into stardom. The requirements of celebrity Chefs are huge: they must be very talented in the kitchen, entrepreneurial both inside and especially outside of the restaurant, highly experienced at marketing and public relations, creative, savvy about business, media-trained, and very personable. With such high require-ments, it is obvious that you need many years of experience before you can even consider approaching the Food Network or cookbook publisher, or open-ing more restaurants. In any case, few Chefs ever become celebrity Chefs, and indeed, many Chefs are not interested in pursuing celebrity status. Chefs, such as Michael Romano of New York's Union Square Café, would rather run his own restaurant and make wonderful food for a smaller audience.

## ADVICE FROM A RECENT GRADUATE

Name:       Adam Volk
Education:  AOS Culinary Arts, Culinary Institute of America, 2006

*[G]et to know your Chef Instructors. They are an invaluable resource.*

## EXERCISES

1. Learn about the field of business communications at the website of the International Association of Business Communications (www.iabc.com). On their home page, click on "About IABC" in the pull-down box. Which of the job positions discussed there are found in hotels or restaurant chains, managed services companies, or other large foodservices? What does each position do? Do any interest you?

2. Many food stylists and food photographers have their own websites, which they use as a portfolio. Find and critique a website for a food stylist or photographer.

3. Find out if there are personal Chefs in your area. Use the phone book or the Internet, or just ask around.

4. Would you like to work in any of the jobs discussed in this chapter? Why or why not?

APPENDIX
A
Culinary Professional Organizations

## AMERICAN CULINARY FEDERATION (ACF)

### Who Are They?

The American Culinary Federation (ACF) is the largest and most prestigious organization dedicated to professional Chefs in the United States today. Their mission is as follows.

It is our goal to make a positive difference for culinarians through education, apprenticeship, and certification, while creating a fraternal bond of respect and integrity among culinarians everywhere.

ACF offers its members many opportunities to keep up to date with the latest in knowledge and skills through its journals, seminars, workshops, national conventions, and regional conferences. ACF also sanctions U.S. culinary competitions and oversees international competitions that take place in the United States. ACF accredits culinary programs at the secondary and postsecondary levels. Local chapters of ACF offer members opportunities to network with nearby culinary professionals. As a member of ACF, you are simultaneously enrolled in the World Association of Chefs' Societies (WACS), which represents 54 countries.

ACF offers the only comprehensive certification program for Chefs in the United States. ACF certification is a valuable credential awarded to Cooks and Chefs and Pastry Cooks after a rigorous evaluation of industrial experience, professional education, and detailed testing.

### Where Are They?

180 Center Place Way
St. Augustine, FL 32095
800-624-9458
www.acfchefs.org

PUBLICATIONS
Monthly magazine: *The National Culinary Review*
Monthly newsletter: *Center of the Plate*

CERTIFICATIONS
Certified Master Chef (CMC)
Certified Master Pastry Chef (CMPC)
Certified Executive Chef (CEC)
Certified Chef de Cuisine (CCC)
Certified Executive Pastry Chef (CEPC)
Personal Certified Executive Chef (PCEC)
Certified Culinary Administrator (CCA)
Certified Culinary Educator (CCE)
Certified Secondary Culinary Educator (CSCE)
Certified Working Pastry Chef (CWPC)
Certified Sous Chef (CSC)
Personal Certified Chef (PCC)
Certified Pastry Culinarian (CPC)
Certified Culinarian (CC)

## AMERICAN DIETETIC ASSOCIATION (ADA)

### Who Are They?

The American Dietetic Association (ADA) is the largest and most visible group of professionals in the nutrition field. Most members are Registered Dietitians (RDs). Individuals with the RD credential have specialized education in human anatomy and physiology, chemistry, medical nutrition therapy, foods and food science, the behavioral sciences, and foodservice management. Registered dietitians must complete at least a bachelor's degree at an accredited college or university, a program of college-level dietetics courses accredited by the Commission on Accreditation for Dietetics Education, a supervised practice experience, and a qualifying examination. Continuing education is required to maintain RD status. Most RDs are licensed or certified by the state in which they live. They work in hospitals and other health-care settings, private practice, sales, marketing, research, government, restaurants, fitness, and food companies.

### Where Are They?

120 South Riverside Plaza, Suite 2000
Chicago, IL 60606
800-877-1600
www.eatright.org

PUBLICATIONS
Monthly journal: *Journal of the American Dietetic Association*
Bimonthly newsletter: *ADA Times*

REGISTRATIONS
Registered Dietitian (RD)
Registered Dietetic Technician (DTR)

## American Hotel & Lodging Association (AH&LA)

### EDUCATIONAL INSTITUTE OF THE AMERICAN HOTEL & LODGING ASSOCIATION (EI)

### Who Are They?

The American Hotel & Lodging Association (AHLA) is a 93-year-old federation of state hotel lodging associations, representing about 11,000 property members worldwide. AH&LA provides its members with assistance in operations, education, and communications, and lobbying in Washington, DC, to encourage a climate in which hotels prosper.

The Educational Institute of the AH&LA (EI) is the leading source of quality hospitality education, training, and professional certification for hospitality schools and colleges and industries around the world.

**Where Are They?**

American Hotel & Lodging Association
1201 New York Avenue NW #600
Washington, DC 20005
202-289-3100
www.ahla.com
American Hotel & Lodging Association Educational
Institute
800 North Magnolia Avenue, Suite 1800
Orlando, FL 32803
800-752-4567
American Hotel & Lodging Association Educational
Institute
2113 North High Street
Lansing, MI 48906
517-372-8800
www.ei-ahla.org

PUBLICATION
Monthly magazine: *Lodging*

CERTIFICATION
The Educational Institute offers a number of professional certifications, including Certified Food and Beverage Executive. The food and beverage manager must be an expert at providing quality service and have excellent leadership and organizational skills, technical proficiency, and a commitment to high standards.
Certification lasts five years, during which requirements for recertification must be fulfilled. Certificate holders must maintain a qualifying position within the industry and earn points by carrying out various activities. The categories in which points are awarded include:

> Professional experience
> Professional development activities/seminars
> Industry involvement
> Educational services

EI provides a portfolio to help track professional development activities.

## AMERICAN INSTITUTE OF BAKING (AIB)

### Who Are They?

The American Institute of Baking (AIB) provides educational and other programs, products, and services to baking and general food production industries around the world. Its members range from international food ingredient and foodservice companies to small single-unit traditional and artisan retail bakeries. AIB is a resource for bakers looking for information and expertise in baking production, experimental baking, cereal science, nutrition, food safety and hygiene, occupational safety, and maintenance engineering.

**Where Are They?**

1213 Bakers Way
P.O. Box 3999
Manhattan, KS 66505-3999
800-633-5137
www.aibonline.org

PUBLICATIONS
Monthly magazine: *AIB Research Technical Bulletin*
Bimonthly magazine: *AIB Maintenance Engineering Bulletin*
Quarterly newsletter: *Bakers Way*

CERTIFICATIONS
Certified Baker — Bread and Rolls
Certified Baker — Cakes and Sweet Goods
Certified Baker — Cookies and Crackers

## AMERICAN INSTITUTE OF WINE & FOOD (AIWF)

### Who Are They?

The American Institute of Wine & Food (AIWF) is an organization devoted to improving the appreciation, understanding, and accessibility of food and drink. With over 30 U.S. chapters, its members include restaurateurs, food industry professionals, food educators, nutritionists, Chefs, wine professionals, and dedicated food and wine enthusiasts.

### Where Are They?

213-37 39th Avenue, Box 216
Bayside, NY 11361
800-274-2493
www.aiwf.org
PUBLICATION
Bimonthly newsletter: *American Wine and Food*

## AMERICAN PERSONAL & PRIVATE CHEF ASSOCIATION (APPCA)

### Who Are They?

The American Personal & Private Chef Association (APPCA) has been training, supporting, and representing successful personal Chefs since 1995.

### Where Are They?

4572 Delaware Street
San Diego, CA 92116
800-644-8389
www.personalchef.com

CERTIFICATIONS (through APPCA and ACF)
Personal Certified Executive Chef (PCEC)
Personal Certified Chef (PCC)

## AMERICAN SOCIETY FOR HEALTHCARE FOOD SERVICE ADMINISTRATORS (ASHFSA)

### Who Are They?

The American Society for Healthcare Food Service Administrators is an affiliate of the American Hospital Association. Members include food and nutrition service management professionals in hospitals, continuing care retirement communities, nursing homes, and other healthcare facilities. Members include Director of Food and Nutrition Services, Director of Dining Services, café/catering/vending managers, clinical nutrition managers, and dietitians. ASHFSA welcomes food and nutrition service professionals from both independent and contract operations.

### Where Are They?

304 West Liberty Street, Suite 201
Louisville, KY 40202
800-620-6422
www.ashfsa.org

PUBLICATION
Quarterly magazine: *Healthcare Food Service Trends*

CERTIFICATIONS/RECOGNITIONS
ASHFSA has a professional recognition program called APEX, which stands for Actions for Professional Excellence. The three levels of achievement are:

Level 1. Accomplished Health Care Foodservice Administrator (AHCFA)
Level 2. Distinguished Health Care Foodservice Administrator (DHCFA)
Level 3. Fellow Health Care Foodservice Administrator (FHCFA)

The ASHFSA Professional Recognition Program is designed to recognize those factors that are indispensable to true professionalism: basic and continuing education, experience, and participation in professional and society activities. Levels of recognition are achieved successively. Each level requires additional education, work experience, and participation in society activities. Members may put the acronym for the level they have accomplished after their name (for example, John Hall, DHCFA).

## ASSOCIATION FOR CAREER AND TECHNICAL EDUCATION (ACTE)

### Who Are They?

The Association for Career and Technical Education (ACTE) is dedicated to the advancement of education that prepares youth and adults for successful careers. ACTE provides resources to enhance the job performance and satisfaction of its members, increases public awareness of career and technical programs, and works on growth in funding for these programs. Its members include over 30,000 teachers, counselors, and administrators at the middle, secondary, and postsecondary school levels.

### Where Are They?

1410 King Street
Alexandria, VA 22314
800-826-9972
www.acteonline.org

PUBLICATIONS
Monthly magazine: *Techniques*
E-mail newsletter: *Career Tech Update*

## BCA (FORMERLY BLACK CULINARIAN ALLIANCE)

### Who Are They?

BCA is an educational and networking association of hospitality and foodservice professionals founded in 1993.

### Where Are They?

55 West 116th Street, Suite 234
New York, NY 10026
800-308-8188
www.blackculinarians.com

## BREAD BAKERS GUILD OF AMERICA

### Who Are They?

The Bread Bakers Guild of America provides education resources for members and fosters the growth of artisan baking and the production of high-quality bread products. Its members include professional bakers, baking educators, home bakers, vendors, and others who share common goals.

### Where Are They?

3203 Maryland Avenue
North Versailles, PA 15137-1629
412-823-2080
www.bbga.org

PUBLICATION
Newsletter: *Bread Bakers Guild*

## CLUB MANAGERS ASSOCIATION OF AMERICA (CMAA)

### Who Are They?

Members of the Club Managers Association of America (CMAA) manage more than 3,000 country, city, athletic, faculty, yacht, town, and military clubs. CMAA provides its members with professional development programs, networking opportunities, publications, and certification programs.

**Where Are They?**

1733 King Street
Alexandria, VA 22314-2720
703-739-9500
www.cmaa.org

PUBLICATION
Monthly magazine: *Club Management*

CERTIFICATIONS
Certified Club Manager (CCM)
Master Club Manager (MCM)

## CONFRÉRIE DE LA CHAÎNE DES RÔTISSEURS

**Who Are They?**

The Confrérie de la Chaîne des Rôtisseurs brings together professional and amateur gastronomes in an organization that celebrates the pleasures of food, wine, and spirits and encourages the development of young professionals. The organization is based on the traditions and standards of the medieval French guild of rôtisseurs, or "meat roasters." Members are in more than 70 countries around the world. The United States has its own chapter.

**Where Are They?**

Chaîne House
Fairleigh Dickinson University
285 Madison Avenue
Madison, NJ 07940-1099
973-360-9200
www.chaineus.org

PUBLICATION
Three times yearly: *Gastronome*

## DIETARY MANAGERS ASSOCIATION (DMA)

**Who Are They?**

The Dietary Managers Association is a national association with over 15,000 professionals dedicated to the mission of "providing optimum nutritional care through food service management." Dietary managers work in nursing homes and other long-term care facilities, hospitals, schools, correctional facilities, and other settings. Responsibilities may include directing and controlling menu planning, food purchasing, food production and service, financial management, employee hiring and training, supervision, nutritional assessment, and clinical care. Dietary managers who have earned the Certified Dietary Manager (CDM), Certified Food Protection Professional (CFPP) credential are also specially trained in food safety and sanitation. Dietary managers may work as foodservice directors, assistant foodservice directors, supervisors, and in other positions.

**Where Are They?**

406 Surrey Woods Drive
St. Charles, IL 60174
800-323-1908
www.dmaonline.org

PUBLICATION
Monthly magazine: *Dietary Manager*

CERTIFICATIONS
Certified Dietary Manager (CDM)
Certified Food Protection Professional (CFPP)

## FOODSERVICE CONSULTANTS SOCIETY INTERNATIONAL (FCSI)

**Who Are They?**

The Foodservice Consultants Society International (FCSI) promotes professionalism in foodservice and hospitality consulting and helps its members by supplying networking and educational opportunities, professional recognition, and other services. FCSI has members who work in layout and design, planning, research, training, technology, operations, and management.

**Where Are They?**

304 West Liberty Street, Suite 201
Louisville, KY 40202
502-583-3783
www.fcsi.org

PUBLICATIONS
Quarterly magazine: *The Consultant*
Monthly e-mail newsletter: *The Forum*

## FOODSERVICE EDUCATORS NETWORK INTERNATIONAL (FENI)

**Who Are They?**

Foodservice Educators Network International (FENI) is a group of foodservice educators who work in high schools and postsecondary programs. FENI works with educators to help them advance their professional growth. A key element of FENI is to facilitate the means for culinary educators to share teaching techniques and other information with colleagues and industry partners to enhance high standards of culinary education.

**Where Are They?**

20 West Kinzie, 12th Floor
Chicago, IL 60610
312-849-2220
www.feni.org

PUBLICATION
Quarterly magazine: *Chef Educator Today*

CERTIFICATION
Certified Culinary Instructor (CCI)

## INSTITUTE OF FOOD TECHNOLOGISTS (IFT)

### Who Are They?

The Institute of Food Technologists (IFT) has members who work in food science, food technology, and related professions in industry, academia, and government.

### Where Are They?

525 West Van Buren, Suite 1000
Chicago, IL 60607
312-782-8424
www.ift.org

### PUBLICATION

Monthly magazine: *Food Technology*

## INTERNATIONAL ASSOCIATION OF CULINARY PROFESSIONALS (IACP)

### Who Are They?

The International Association of Culinary Professionals (IACP) is a group of approximately 4,000 food professionals from over 35 countries. IACP provides continuing education, networking, and information exchange for its members, who work in culinary education, communication, or the preparation of food and drink. Its mission is to "help its members achieve career success ethically, responsibly, and professionally." Many of its members are cooking school instructors, food writers, cookbook authors, Chefs, and food stylists.

### Where Are They?

304 West Liberty Street, Suite 201
Louisville, KY 40202
800-928-4227
www.iacp.com

### PUBLICATION

Quarterly magazine: *Food Forum*

### CERTIFICATIONS

Certified Culinary Professional (CCP)

## INTERNATIONAL CATERERS ASSOCIATION (ICA)

### Who Are They?

The International Caterers Association (ICA) includes both off-premises and on-premises caterers from around the world. ICA provides education, mentoring, and other services for professional caterers, and promotes the profession of catering to the public, vendors, and others.

### Where Are They?

91 Timberlane Drive
Williamsville, NY 14221
877-422-4221
www.icacater.org

### PUBLICATIONS

Bimonthly newsletter: *CommuniCater*
Membership includes complimentary subscriptions to *Catering Magazine*, *Event Solutions*, and *Special Events Magazine*.

## CRUISE LINES INTERNATIONAL ASSOCIATION, INC.

### Who Are They?

Cruise Lines International Association is the world's largest cruise association and is dedicated to the promotion and growth of the cruise industry. CLIA is composed of 24 of the major cruise lines serving North America.

### Where Are They?

2111 Wilson Boulevard, 8th Floor
Arlington, VA 22201
703-522-8463
www.cruising.org

### PUBLICATIONS

Monthly e-mail: *Fast Facts*
Quarterly newsletter: *Even Keel*

## INTERNATIONAL COUNCIL ON HOTEL, RESTAURANT, AND INSTITUTIONAL EDUCATION (I-CHRIE)

### Who Are They?

I-CHRIE is the advocate for schools, colleges, and universities that offer programs in hotel and restaurant management, foodservice management, and culinary arts.

### Where Are They?

2613 North Parham Road
Richmond, VA 23294
804-346-4800
www.chrie.org

### PUBLICATION

Monthly journal: *Journal of Hospitality and Tourism Education*

## INTERNATIONAL FLIGHT SERVICES ASSOCIATION (IFSA)

### Who Are They?

The International Flight Services Association (IFSA) serves the needs and interests of airline and railway personnel as well as airline and rail caterers who are responsible for providing passenger foodservice on regularly scheduled travel routes. IFSA's membership is dedicated to the advancement of the art and science of this segment.

**Where Are They?**

1100 Johnson Ferry Road, Suite 300
Atlanta, GA 30342
404-252-3663
www.ifsanet.com

PUBLICATION
Monthly e-mail newsletter: *Onboard IFSA*

## INTERNATIONAL FOOD SERVICE EXECUTIVES ASSOCIATION (IFSEA)

**Who Are They?**

The International Food Service Executives Association (IFSEA) offers its diverse membership opportunities for personal development, networking, mentoring, and community service. IFSEA's membership includes professionals from many areas of the foodservice industry, including the military.

**Where Are They?**

304 West Liberty Street, Suite 201
Louisville, KY 40202
502-583-3783
www.ifsea.com

PUBLICATION
Magazine: *Hotline*

CERTIFICATIONS
Master Certified Food Executive (MCFE)
Certified Food Executive (CFE)
Certified Food Manager (CFM)

## INTERNATIONAL FOODSERVICE MANUFACTURERS ASSOCIATION (IFMA)

**Who Are They?**

The International Foodservice Manufacturers Association (IFMA) represents manufacturers of foodservice equipment. IFMA provides its members with many services, including training, networking, visibility, and marketing information.

**Where Are They?**

Two Prudential Plaza
180 North Stetson Avenue, Suite 4400
Chicago, IL 60601
312-540-4400
www.ifmaworld.com

PUBLICATION
Bimonthly magazine: *IFMA World*

## LES DAMES D'ESCOFFIER

**Who Are They?**

Les Dames D'Escoffier is an invitation-only culinary organization of successful women leaders.

The organization mentors young women, educates the public about the pleasures of the table, awards scholarships, and supports food-related charities. Members include Chefs, restaurateurs, cookbook authors, food journalists and historians, wine professionals, food publicists, culinary educators, and hospitality executives.

**Where Are They?**

212-867-3929
www.ldei.org

## MILITARY HOSPITALITY ALLIANCE (MHA)

**Who Are They?**

The Military Hospitality Alliance (MHA) is the affiliate of the International Food Service Executives Association that focuses on the needs of the military. Projects include a military culinary competition, Enlisted Aide of the Year awards, and more.

**Where Are They?**

836 San Bruno Avenue
Henderson, NV 89015
888-234-3732
www.mhaifsea.com

CERTIFICATIONS
Registered Military Culinarian (RMC)
Registered Military Hospitality Manager (RMHM)

## NATIONAL ASSOCIATION FOR THE SPECIALTY FOOD TRADE (NASFT)

**Who Are They?**

The National Association for the Specialty Food Trade (NASFT) is an international organization of domestic and foreign manufacturers, importers, distributors, brokers, retailers, restaurateurs, caterers, and others in the specialty foods business.

**Where Are They?**

120 Wall Street, 27th Floor
New York, NY 10005
212-482-6440
www.specialtyfood.com

PUBLICATION
Monthly magazine: *Specialty Food*

## NATIONAL ASSOCIATION OF COLLEGE & UNIVERSITY FOOD SERVICES (NACUFS)

**Who Are They?**

The National Association of College & University Food Services (NACUFS) is the trade association for foodservice professionals at nearly 650 institutions of higher education in the United States, Canada, and abroad.

**Where Are They?**

2525 Jolly Road, Suite 380
Okemos, MI 48864-3680
517-332-2494
Fax: 517-332-8144
www.nacufs.org

PUBLICATION
Quarterly magazine: *Campus Dining Today*

## NATIONAL FOOD PROCESSORS ASSOCIATION (NFPA)

**Who Are They?**

The National Food Processors Association provides food-science and technical expertise for the food-processing industry on topics such as food safety, nutrition, and technical and regulatory matters. NFPA members process and package fruit, vegetables, meat, fish, and other foods.

**Where Are They?**

1350 I Street NW, Suite 300
Washington, DC 20005
800-355-0983

PUBLICATION
Monthly journal: *NFPA Journal*

## NATIONAL ICE CARVING ASSOCIATION (NICA)

**Who Are They?**

The National Ice Carving Association (NICA) is the only organization in the United States devoted solely to promoting the art of ice sculpture across the country and around the world. With nearly 500 members, NICA sanctions and organizes ice-carving competitions in North America, providing standardized guidelines for judging the quality of ice sculptures in competition.

**Where Are They?**

P.O. Box 3593
Oak Brook, IL 60522-3593
630-871-8431
www.nica.org

PUBLICATION
Newsletter: *On Ice*

CERTIFICATIONS
Certified Master Ice Carver
Certified Professional Ice Carver
Certified Competition Ice Carver
Certified Ice Carver

## NATIONAL RESTAURANT ASSOCIATION (NRA)

## THE NATIONAL RESTAURANT ASSOCIATION EDUCATIONAL FOUNDATION (NRAEF)

**Who Are They?**

The National Restaurant Association (NRA) is the leading business association for the restaurant industry.

Together with the National Restaurant Association Educational Foundation, the Association's mission is to represent, promote, and educate (through reports, publications, research, training materials, networking, etc.) the restaurant and foodservice industry.

The National Restaurant Association Educational Foundation (NRAEF) is the not-for-profit organization dedicated to fulfilling the educational mission of the National Restaurant Association. The NRAEF is the premier provider of education resources, materials, and programs to attract, develop, and retain the industry's workforce. Examples of NRAEF programs include ServSafe food safety certification, ServSafe Alcohol certification, the ProStart School-to-Career program, the ManageFirst program, and the Foodservice Management Professional (FMP) certification program.

**Where Are They?**

National Restaurant Association
1200 17th Street NW
Washington, DC 20036
202-331-5900
www.restaurant.org
National Restaurant Association Educational Foundation
175 West Jackson Boulevard, Suite 1500
Chicago, IL 60604-2814
800-765-2122
312-715-1010 (in Chicago)

PUBLICATION (through the National Restaurant Association)
Monthly online magazine: *Restaurants USA Online*

CERTIFICATION (through Educational Foundation)
Foodservice Management Professional (FMP)

## NATIONAL SOCIETY FOR HEALTHCARE FOODSERVICE MANAGEMENT (HFM)

**Who Are They?**

The National Society for Healthcare Foodservice Management (HFM) is a professional association representing healthcare foodservice operators and their suppliers. HFM accepts only members who operate independent operations and are not contracted. HFM offers advocacy for independent healthcare foodservices as well as management tools to decrease costs, increase patient and staff satisfaction, and define successful operational performance. Members are mostly from hospitals.

**Where Are They?**

355 Lexington Avenue, 15th Floor
New York, NY 10017
212-297-2166
www.hfm.org

PUBLICATION
Monthly magazine: *Innovator*

## NORTH AMERICAN ASSOCIATION OF FOOD EQUIPMENT MANUFACTURERS (NAFEM)

**Who Are They?**

The North American Association of Food Equipment Manufacturers (NAFEM) represents companies throughout the United States, Canada, and Mexico that manufacture commercial foodservice equipment and supplies.

**Where Are They?**

161 North Clark Street, Suite 2020
Chicago, IL 60601
312-821-0201
www.nafem.org

PUBLICATION
Quarterly magazine: *NAFEM in Print*

CERTIFICATION
Certified Food Service Professional (CFSP)

## RESEARCH CHEFS ASSOCIATION (RCA)

**Who Are They?**

The Research Chefs Association (RCA) brings together chefs, food scientists, and others who work in food research and development in restaurants, food companies, and many other businesses.

**Where Are They?**

1100 Johnson Ferry Road, Suite 300
Atlanta, GA 30342
404-252-3663
www.culinology.com

PUBLICATIONS
Magazine: *Culinology*
Quarterly newsletter: *Culinology Currents*

CERTIFICATIONS
Certified Research Chef (CRC)
Certified Culinary Scientist (CCS)

## RETAIL BAKERS OF AMERICA (RBA)

**Who Are They?**

The Retail Bakers of America (RBA) is a trade association of independent retail bakeries. The association works to improve the operations and profitability of its members by offering training, networking, conventions, communications, and meetings.

**Where Are They?**

8201 Greensboro Drive, Suite 300 McLean, VA 22102
**Telephone**: 703.610.9035
www.rbanet.com

CERTIFICATIONS
Certified Journey Baker (CJB)
Certified Baker (CB)
Certified Decorator (CD)
Certified Bread Baker (CBB)
Certified Master Baker (CMB)

## SCHOOL NUTRITION ASSOCIATION (FORMERLY THE AMERICAN SCHOOL FOOD SERVICE ASSOCIATION)

**Who Are They?**

The School Nutrition Association (SNA) has over 55,000 members involved in some way with the National School Lunch Program. SNA works on making sure all children have access to healthful, tasty school meals and nutrition education. SNA does this by providing education and training, setting standards, and educating members on legislative, industry, nutritional, and other issues.

**Where Are They?**

700 South Washington Street, Suite 300
Alexandria, VA 22314
703-739-3900
www.schoolnutrition.org

PUBLICATIONS
Monthly magazine: *School Foodservice and Nutrition*
Semiannual journal: *Journal of Child Nutrition and Management*

CREDENTIAL
School Foodservice and Nutrition Specialist (SFNS)

## SOCIÉTÉ CULINAIRE PHILANTHROPIQUE

**Who Are They?**

The Société Culinaire Philanthropique is the oldest association of Chefs and Cooks in the United States. Founded by a group of French Chefs, its members organize the annual Salon of Culinary Arts in New York City.

**Where Are They?**

305 East 47th Street, Suite 11B
New York, NY 10017
212-308-0628
www.societeculinaire.com

## SOCIETY FOR FOODSERVICE MANAGEMENT (SFM)

**Who Are They?**

The Society for Foodservice Management (SFM) serves the needs and interests of executives in the on-site foodservice industry (predominantly B&I). SFM provides member interaction, continuing education, and professional development via information and research.

**Where Are They?**

304 West Liberty Street, Suite 201
Louisville, KY 40202
502-583-3783
www.sfm-online.org

PUBLICATION
Monthly e-mail newsletter: *FastFacts*

## UNITED STATES PERSONAL CHEF ASSOCIATION (USPCA)

**Who Are They?**

The United States Personal Chef Association serves the needs and interests of personal Chefs. USPCA offers training, certification, personal Chef software, and other services.

**Where Are They?**

610 Quantum Road, NE
Rio Rancho, NM 87124
800-995-2138
www.uspca.com

PUBLICATION
Monthly magazine: *Personal Chef*

CERTIFICATION
Certified Personal Chef (CPC)

## WOMEN CHEFS & RESTAURATEURS (WCR)

**Who Are They?**

The Women Chefs & Restaurateurs (WCR) promote the education and advancement of women in the restaurant industry. Their goals are "exchange, education, enhancement, equality, empowerment, entitlement, environment, and excellence." WCR offers a variety of networking, professional, and support services.

**Where Are They?**

304 West Liberty Street, Suite 201
Louisville, KY 40202
877-927-7787
www.womenchefs.org

PUBLICATION
Quarterly newsletter: *Entrez!*

## WOMEN'S FOODSERVICE FORUM (WFF)

**Who Are They?**

The Women's Foodservice Forum was founded in 1989 to develop leadership talent and ensure career advancement of executive women in the foodservice industry. WFF currently offers three types of mentor programs and many other resources to help women move into senior-level positions. Its membership comes from restaurant operations, manufacturing, distribution, publishing, and consulting.

**Where Are They?**

1650 W 82nd Street, Suite 650
Bloomington, MN 55431
866-368-8008
www.womensfoodserviceforum.com

PUBLICATION
Monthly newsletter: *Open Doors*

# Job Descriptions

## RESTAURANT EXECUTIVE CHEF

**Position Type:** Full-Time/Permanent

**Pay Basis:** Salary

**Reports to:** General Manager

### RESPONSIBILITIES:

#### Menu Planning
Develops dining room menus and revises twice annually

Develops catering menu and customizes menu as necessary

#### Purchasing and Inventory
Sets appropriate purchasing specifications

Purchases all food and supplies and ensures items are available on a timely basis

Obtains fair pricing by using the bidding process

Completes all purchasing and inventory records within guidelines

Maintains inventory controls

#### Food Production and Presentation
Accurately forecasts food production needs

Completes all food production records within guidelines

Coordinates activities of cooks and other kitchen personnel engaged in preparing and cooking foods

*(Continued)*

Prepares food as needed

Develops and standardizes production recipes to ensure consistent quality

Develops quality standards for food quality and presentation

Checks food quality, presentation, and portion control before meals

Ensures maintenance of cooking equipment

## Sanitation

Maintains a sanitary and safe kitchen/storage environment for all employees

Implements HACCP plan

## Financial

Manages/controls food costs

Prepares budgets in consultation with General Manager

## Supervision/Human Resources

Supervises and motivates cooking staff

Interviews and hires cooking staff

Trains cooking staff

Evaluates employee performance

## Leadership

Serves as a role model for employees through teamwork support, attention to detail, quality control, and interactions with guests and team members

Ensures guest satisfaction

Ensures that employees receive fair and consistent treatment and opportunities to advance

## QUALIFICATIONS:

Bachelor's or related culinary degree required

7+ years experience as a Chef or Sous Chef required

ACF-Certified Executive Chef certification preferred

Food handler's certification and knowledge of HACCP standards required

Skilled in menu development within budgeted food costs

Knowledge about plating and presentation

Experience in supervising and training staff

Knowledge of Word and Excel

Strong oral and written communication skills

# RESTAURANT PASTRY CHEF

## SUMMARY:

Under limited supervision, oversees food and pastry production for all food outlets and banquet functions. Develops menus, food purchase specifications, and recipes. Directly supervises all production and pastry staff. Maintains highest professional food quality and sanitation standards.

## DUTIES AND RESPONSIBILITIES:

1. Hires, trains, and supervises the work of food and pastry production staff
2. Plans menus for all foodservice locations, considering customer base, popularity of dishes, holidays, costs, and a wide variety of other factors
3. Schedules and coordinates the work of chefs, cooks, and other kitchen employees to ensure that food preparation is economical and technically correct
4. Conducts regular physical inventories of food supplies and assesses projected needs; orders all food and supplies for catering and cash operations
5. Ensures that high standards of sanitation and cleanliness are maintained throughout the kitchen areas at all times
6. Establishes controls to minimize food and supply waste and theft
7. Safeguards all food preparation employees by implementing training to increase their knowledge of safety, sanitation, and accident prevention principles
8. Develops and tests recipes and techniques for food preparation and presentation that help ensure consistent high quality and minimize food costs; exercises portion control over all items served and assists in establishing menu selling prices
9. Prepares necessary data for the budget in area of responsibility; projects annual food and labor costs and monitors actual financial results; takes corrective action where necessary to help ensure that financial goals are met
10. Consults with catering staff about food production aspects of special events being planned
11. Cooks or directly supervises the cooking of items that require skillful preparation
12. Evaluates food products to ensure that quality standards are consistently attained
13. Performs miscellaneous job-related duties as assigned

## MINIMUM JOB REQUIREMENTS:

Certificate or college degree from accredited culinary school

State foodservice safe food handler's certificate

American Culinary Federation certification preferred

Five to 8 years of experience with some supervisory experience

*(Continued)*

## CONDITIONS OF EMPLOYMENT:

Successful candidate must submit to postoffer, preemployment physical examination/medical history check.

## WORKING CONDITIONS AND PHYSICAL EFFORT:

Work involves moderate exposure to unusual elements such as extreme temperatures, dirt, dust, fumes, smoke, unpleasant odors, and loud noises.

Moderate physical activity. Requires handling of average-weight objects up to 25 pounds or standing and/or walking for more than four (4) hours per day.

Work environment involves some exposure to hazards or physical risks, which requires following basic safety precautions.

# HOTEL EXECUTIVE CHEF

## DIRECT REPORT

Food and Beverage Director or General Manager

## MISSION

To achieve food and beverage revenue, profit, and customer satisfaction goals by developing and implementing a menu that creates a luxury dining experience

## POINTS OF INTERACTION:

Guests

Employees

Hotel owners

Management team

The local community

Vendors

## SUCCESS FACTORS:

Customer loyalty and satisfaction

Meeting or exceeding food and beverage revenue, profit, and cost goals

Maintaining employee satisfaction and retention goals

Ensuring a constant adherence to all quality standards and operational procedures

## RESPONSIBILITIES:

### Financial

Develops and implements strategies for the kitchen that support achievement of goals

Develops and implements business strategies for hotel that are aligned with overall mission, vision, values, and strategies

Monitors status regularly and adjusts strategies as appropriate

Develops the annual budget and participates in forecasting in conjunction with the controller and the Food and Beverage Director

Creates an environment for employees that is aligned with the hotel culture through constant communication and reinforcement

### Guest/Employee Experience

Communicates the service vision to, and reinforces it for, managers and employees

Creates an environment designed to stimulate all senses through personal services, amenities, and experiences provided by employees

Ensures that the hotel delivers the desired experience by reviewing hotel operations from the customer's perspective, as well as from a business perspective

Keeps current on pulse of the guests, constantly seeking opportunities to follow up on their experience

Provides employees with the tools and environment they need to meet desired job performance goals

Develops and implements strategies to support employee enrichment through creative learning programs and mentoring

### Leadership

Researches customer preferences and develops a menu that is marketable to the customer base

Creates the environment outlined by the hotel's business plan

Incorporates regional flavors and methodologies that reflect the local area

Instigates the culture that drives the kitchen philosophy and motivates employees

Serves as a leadership role model for every employee and manager, fostering teamwork, attention to detail, quality control, and care for all facets of departmental operation

Exhibits above-average personal standards in all areas related to the position

Identifies, develops, and implements business projects through careful planning and detailed guidelines

Demonstrates the ability to relate to, communicate with, and motivate team members to sustain acceptable performance

### Culinary Operation

Coordinates service with restaurant and banquet operations

Maintains product consistency by inspecting seasonings, portioning, and appearance of food

Seeks out sources for quality product

*(Continued)*

Monitors all perishables and meat for freshness

Develops menu pricing structure and adherence to usage guidelines

Oversees the inventory, purchasing, and disbursement of all supplies

Ensures that proper sanitation practices are followed

Recruits new staff and participates in staff selection

Directs staff scheduling based on forecasted volumes

Ensures that all controls and procedures are in compliance with policy

## QUALIFICATIONS:

Three+ years senior-level Executive Sous Chef/Executive Chef experience in high-quality volume-oriented environment

AOS Culinary School graduate required, bachelor's degree preferred

Ability to handle multiple tasks and work well in environment with time constraints

Excellent communication skills, both oral and written

Ability to develop, implement, and maintain systems

Ability to generate new ideas and implement them

Effective leadership skills

State food handler's certificate

Must be detail-oriented and able to perform work accurately and on time

Ability to meet deadlines and work independently

Ability to work flexible work schedule

# HOTEL PASTRY CHEF

## ROLE DESCRIPTION

**Reports to:** Executive Sous Chef or Executive Chef

**Purpose:** To achieve food and beverage revenue, profit, and customer satisfaction goals by managing the pastry department for a five-star dining experience

## CONSTITUENTS:

Guests

Employees

Management team

The local community

Vendors

### Indicators of Success

Achievement of customer satisfaction and loyalty goals

Quality of food product

Achievement of employee satisfaction and retention goals

Achievement of the hotel's financial goals

### Areas of Responsibility

Participates in the development and implementation of business strategies for the kitchen that are aligned with the hotel's overall mission, vision, values, and strategies

Participates in the development of the kitchen's business strategies

Creates menu for the pastry department

Creates an environment at the hotel designed to stimulate all senses through personal service, amenities, and experiences provided by employees

Ensures that the hotel delivers the intended experience by reviewing hotel operations from the customer's perspective, as well as from a business perspective

Keeps current on the pulse of the guests, constantly seeking opportunities to follow up on their experiences

Provides employees with the tools and environment they need to perform

Develops and implements strategies and practices that support employee engagement

Creates luxury for the senses by managing the preparation of baked goods, pastries, and desserts for a fine hotel restaurant

Develops menu for baked goods and desserts

Creates signature dessert items

Creates daily specials based on availability of ingredients

Recruits, selects, and schedules staff

Develops and implements training plan for employees that includes culinary skills, customer service skills, and kitchen safety

Ensures adherence to standards of food quality

Maintains product consistency by inspecting seasonings, portioning, and appearance of food

Ensures that unused food is stored properly in order to minimize waste and maximize quality

Ensures that proper sanitation practices are followed

Is responsible for bread ordering (with business fluctuation)

Is responsible for inventory and ordering of pastry supplies

Is responsible for all pastry production, including breakfast buffet, à la carte, banquet, and room service

### Analyzes Business Volume and Product Usage Daily

Anticipates business volume by consulting with Executive Chef

Manages inventory of food; provides Executive Chef with recommended stocking levels and purchasing requisitions

*(Continued)*

Keeps current with local competition; continually seeks to improve skills and culinary management expertise

Must be a teacher, able to communicate and show staff new and current ideas

### Success Factors

Attention to Detail—Ensures that data are accurate, work is thorough, and products meet the highest standards

Focus on the Customer—Seeks to understand the (internal/external) customer and meet the needs of both the customer and the company

Drive for Results—Works to achieve high levels of personal and organizational performance in order to meet or exceed objectives

Continuous Improvement—Constantly assesses and adapts current practices to perform a task better, faster, or more efficiently

Information Sharing—Provides information so coworkers, customers, and suppliers understand and can take action

Teamwork—Works well in a team environment and motivates teams to sustain exceptional levels of performance

Development of Self and Others—Continually works to develop own capabilities and those of others

Professionalism and Product or Technical Expertise—Demonstrates the ability to apply technical, professional, or product expertise to real-world situations

Creative Thinking—Develops innovative approaches and imaginative solutions that meet real needs

Strong Relationships—Fosters trust and cooperation among coworkers, customers, and suppliers; develops and sustains personal contact in order to provide mutual benefit

Delegation—Assigns tasks using such techniques as needs analysis, individual skills assessment, objective setting, and communication

Organization—Demonstrates ability to proactively prioritize needs, put first things first, and effectively manage resources

Supervision, Performance Management—Demonstrates ability to relate to, communicate with, and motivate employees to sustain high performance and quality levels

Planning—Shows skill in determining whether tasks should be attempted, identifies the most effective way to complete the task, and prepares for unexpected difficulties

# COUNTRY CLUB CHEF

**Job Title:** Chef

**Position Type:** Full Time/Permanent

**Pay Basis:** Salary + Bonus

## GENERAL DUTIES:

Working Chef for private country club with exclusive clientele. Responsible for all food production, including meals for restaurants, banquets, and other outlets. Develop menus, food purchase specifications, and recipes. Supervise production. Develop and monitor food and labor budget for the department. Maintain highest food quality and sanitation standards.

## SPECIFIC DUTIES:

1. Plans menus for formal dining room, terrace, grill, snack bars, banquet room, and other outlets as requested, considering guests, popularity of various dishes, holidays, costs, and many other factors

2. Hires, trains, and supervises the work of staff in the kitchen and grill

3. Schedules and coordinates the work of Sous Chefs, cooks, and other kitchen employees, so food preparation and presentation are technically correct and cost-effective

4. Develops recipes and plate presentation for high quality, keeping in mind food costs and budget

5. Approves the requisition of food, sanitation supplies, and smallwares

6. Supervises portion control over all items and establishes menu selling prices with Clubhouse Manager

7. Ensures that high standards of sanitation and cleanliness are maintained throughout the areas at all times

8. Trains all cooking personnel on sanitation, food safety, and accident prevention principles

9. Prepares all necessary data for the budget, projects food and labor costs, and monitors actual financial results; takes corrective action where necessary and helps ensure that financial goals are met

10. Attends food and beverage meetings

11. Consults with the event coordinator about food production for special events

12. Personally cooks or directly supervises the cooking of items that require skillful preparation

13. Evaluates food products and interacts with guests to ensure food quality and guest satisfaction

14. Plans and manages the employee meal program

*(Continued)*

**QUALIFICATIONS:**

Self-motivated, organized, well-spoken individual with excellent interpersonal skills. Extensive prior supervisory experience as a Chef required. Prior experience in large private country club preferred. Chef certification preferred.

# BANQUET CHEF

**Position Type:** Full-Time/Permanent

**Pay Basis:** Salary

**Reports To:** Executive Chef

## RESPONSIBILITIES:

Maintains the highest quality of food production and operational standards in order to exceed guest expectations and maximize profits

Supervises all phases of food production and service related to banquet functions

Controls the quality of food production

Controls the quantity of food production so as to avoid excessive leftovers or waste

Supervises banquet kitchen and staff

Communicates with others to plan and coordinate efforts for scheduled functions

Maintains an open and positive working relationship with union members, in agreement with the collective bargaining agreement

Maintains an effective training program for all employees

Does yearly performance evaluations of all employees

Recommends hiring decisions

Maintains sanitation standards

Maintains equipment

Assists in monthly and weekly forecasts

Provides input for annual expense budget and capital budget

## REQUIREMENTS:

Two-year associate's degree or four-year bachelor's degree in culinary

Minimum two years' supervisory experience with proven team-building, problem-solving, and training skills in a large operation

Excellent communication and interpersonal skills

Proven ability to motivate, guide, and coach diverse employees

Strong organizational and time-management skills

Ability to work under stress and in a fast-paced environment

Flexibility with hours, days off, assignments, and additional duties

# CRUISE LINES EXECUTIVE CHEF

**DIRECT REPORT:**

Hotel Manager

Corporate Executive Chef

**POSITIONS REPORTING TO:**

Chef de Parties

Galley Assistants

**ROTATION:**

Two to three months on board vessel; one month vacation

**RESPONSIBILITIES:**

The Executive Chef is responsible for all aspects of the foodservice operation, including the administrative and technical duties required to provide the passengers and crew with high-quality cuisine. He/she is also responsible for supervising the duties of other kitchen personnel. In addition, the Chef ensures that the galley consistently operates under the rules and regulations of the Centers for Disease Control. The Chef maintains communication in all aspects of the foodservice operation, and consistently communicates with the cruise line corporate hotel office personnel.

**MAJOR RESPONSIBILITIES/SPECIFIC TASKS:**

- Keeps the entire foodservice operation in an absolutely sanitary condition per the cruise line and the Centers for Disease Control standards

- Maintains a clear understanding of the company safety and environmental policy and how to access it

- Maintains fluency with accepted rules and regulations set by the Centers for Disease Control and practices and manages these routinely

- Knows the proper use of all kitchen equipment and the proper procedures and detergents used for cleaning and sanitizing all equipment

- Is responsible for daily food production and quality of all menu items served

- Has ability to prepare all standard menu items within production guidelines

- Supervises the loading of all food provisions on turnaround day

- Stays current on developments in the culinary field so updates can be added or changes made to menus with the approval of the corporate Executive Chef

- Maintains and evaluates the Galley Assistant's schedule and work lists in order to maintain sanitation standards

*(Continued)*

- Provides high-quality crew meals, using creativity in utilization of all food products
- Communicates effectively with the Hotel Manager and discusses any matter concerning the operation of the onboard hotel department
- Is responsible for shutting down and securing all ranges and other cooking equipment
- Is responsible for securing and locking all refrigeration/freezer/storage areas
- Is responsible for the general appearance of the Chef de Parties and Galley Assistants
- Maintains an overall awareness of all quality-control functions of the vessel and reports deficiencies to department heads
- Fills in as Chef de Partie whenever necessary
- Secures galley and all galley storage areas in preparation for rough weather
- Attends daily departmental meeting held by the Hotel Manager

### Administrative Duties

- Processes all invoices, control forms, and weekly menu copies on a timely basis and sends them to the corporate office for processing
- Accurately and proficiently follows, completes, receives, and stores scheduled provisioning requirements while staying within budget guidelines for all items ordered
- Refuses or accepts incoming merchandise
- Is responsible for daily menu planning, in line with the guidelines established by the corporate Executive Chef
- Adheres to the concept of utilizing a block or preplanned menu setup, which manages menu flow and passenger expectation; coordinates menu printing and distribution with the Assistant Hotel Manager
- Is responsible for ensuring that all foods are biologically 100% fit for human consumption; is allowed to make menu revisions or substitutions, when necessary, to ensure that the food is maintained at high standards, subject to budget guidelines
- When needed, assists other officers in ordering supplies necessary for the ship to run effectively
- Maintains a positive relationship with all food vendors
- Maintains a positive relationship with all onboard officers

# BUSINESS AND INDUSTRY CHEF

## Chef, Private Executive Dining

**Position Type:** Full-Time/Permanent

**Pay Basis:** Salary

**Reports to:** Executive Chef

### RESPONSIBILITIES:

Supervises and coordinates all related culinary activity for executive dining rooms

Develops daily custom menus for the CEO's office and board events

Develops and trains waitstaff to understand menu concepts

Prepares scratch baked goods for catering and elsewhere

Establishes presentation techniques and quality standards

Plans and prices menus

Estimates food consumption and requisitions food

Selects and develops standardized recipes

Controls inventory

Ensures proper safety and sanitation in kitchen

Ensures proper equipment operation and maintenance

Maintains a commanding presence in the dining room and interacts well with client

Is able to work some nights and weekends

### QUALIFICATIONS:

Bachelor's degree preferred, or related culinary degree

Certified Executive Chef preferred

Seven+ years of culinary management experience

Ability to focus on client services

Experience with food and labor cost control, profit and loss accountability, demonstration cooking, menu development, pricing, and development of culinary team

Contract-managed service experience desirable

# RESEARCH AND DEVELOPMENT CULINARY MANAGER

## Culinary Manager

**Position Type:** Full Time/Permanent

**Pay Basis:** Salary

### SUMMARY

This individual will support the business strategies of the foodservice division by providing expertise and experience in food science, foodservice operations, culinary arts, and product knowledge as an integrated business team member. This individual will build relationships with customers via their culinary teams, in order to identify customer needs and possible solutions. These solutions will be found in the company's current product lines through menu/recipe applications and through new product concepts. This individual will be a communication link with research and development by providing customer insights and jointly working on new product concepts. Also, this individual will provide industry and culinary expertise to photography projects and communications related to food and menu trends, and orchestrate product samples for special events or specific customers. This individual will also be responsible for recipe development for advertising, public relations programs, brochures, and promotions, and will provide culinary presentations to customers and internal groups. Travel will be approximately 25–35%.

### POSITION REQUIREMENTS AND SPECIFICATIONS:

Bachelor's degree in food science, nutrition, or business area strongly preferred; culinary degree required

Minimum of ten years' experience within the foodservice industry, in particular with a variety of national full-service family-style restaurant (FSR) organizations; corporate culinary experience desired

Proven expertise in food science, foodservice operations, and culinary arts as they relate to FSR operations

Strong presentation skills to both internal and external customers

Experience with creative recipe and menu development

Strong planning and consulting skills, specifically to uncover and build new food applications or product lines

Ability to manage multiple projects as well as bring projects to successful completion

Effective time management, communications, and decision-making skills

### IN ADDITION, THE FOLLOWING CRITICAL SUCCESS FACTORS ARE NECESSARY:

Proven passion for food and the desire to build on expertise and industry knowledge

Strong business orientation, with an understanding of the costs and labor involved in product development

*(Continued)*

Confidence in culinary abilities and willingness to share insights and opinions with internal and external customers

Strong team orientation, including the ability to lead a project team and partner with a variety of personalities and styles

Ability to handle multiple projects and adapt to new situations

Customer-service orientation and ability to respond to customer needs

Strong written and oral communication skills, including presentation skills

# ASSISTANT PROFESSOR, CULINARY ARTS

## Culinary Arts Program at Community College Assistant Professor

**Program:** Culinary Arts

**Position Type:** Full Time/Permanent

**Pay Basis:** Salary

**Reports To:** Dean, Business Studies Department

### RESPONSIBILITIES:

1. Teaches coursework within the culinary arts program
2. Shows evidence of planning classes, care in preparation of teaching materials and handouts, and reflective and conscientious feedback on students' work
3. Teaches using a variety of teaching methods that allow students to be actively involved as much as possible; offers learning experiences that help students attain high levels of knowledge, a repertoire of core thinking skills, and culinary skills they will need to draw on in the workplace
4. Fosters understanding of diversity through educational experiences
5. Maintains four office hours each week
6. Recruits students
7. Advises 40 students each year
8. Evaluates transcripts of college students who want to enter one of the culinary programs
9. Establishes and maintains collegial relationships with local hospitality employers and internship sites; secures and maintains contractual agreements
10. Participates in continuing education activities

*(Continued)*

**11.** Promotes high standards of ethical and professional conduct

**12.** Participates in yearly evaluation process, sets yearly goals

## QUALIFICATIONS:

**Education:** Master's degree required

**Experience:** ACF certification required. At least five years' experience as a Chef. Teaching experience preferred.

# Index